Lecture Notes in Mathematics

2019

Editors:
J.-M. Morel, Cachan
B. Teissier, Paris

Subseries:
École d'Été de Probabilités de Saint-Flour

Saint-Flour Probability Summer School

The Saint-Flour volumes are reflections of the courses given at the Saint-Flour Probability Summer School. Founded in 1971, this school is organised every year by the Laboratoire de Mathématiques (CNRS and Université Blaise Pascal, Clermont-Ferrand, France). It is intended for PhD students, teachers and researchers who are interested in probability theory, statistics, and in their applications.

The duration of each school is 13 days (it was 17 days up to 2005), and up to 70 participants can attend it. The aim is to provide, in three high-level courses, a comprehensive study of some fields in probability theory or Statistics. The lecturers are chosen by an international scientific board. The participants themselves also have the opportunity to give short lectures about their research work.

Participants are lodged and work in the same building, a former seminary built in the 18th century in the city of Saint-Flour, at an altitude of 900 m. The pleasant surroundings facilitate scientific discussion and exchange.

The Saint-Flour Probability Summer School is supported by:

– Université Blaise Pascal
– Centre National de la Recherche Scientifique (C.N.R.S.)
– Ministère délégué à l'Enseignement supérieur et à la Recherche

For more information, see back pages of the book and
http://math.univ-bpclermont.fr/stflour/

Jean Picard
Summer School Chairman
Laboratoire de Mathématiques
Université Blaise Pascal
63177 Aubière Cedex
France

Robert J. Adler • Jonathan E. Taylor

Topological Complexity of Smooth Random Functions

École d'Été de Probabilités
de Saint-Flour XXXIX – 2009

 Springer

Robert J. Adler
Faculty of Electrical Engineering
Technion - Israel Institute of Technology
Haifa, 32000
Israel
robert@ee.technion.ac.il

Jonathan E. Taylor
Stanford University
Department of Statistics
Sequoia Hall
390 Serra Mall
Stanford, CA 94305-4065
USA
jonathan.taylor@stanford.edu

ISBN 978-3-642-19579-2   e-ISBN 978-3-642-19580-8
DOI 10.1007/978-3-642-19580-8
Springer Heidelberg Dordrecht London New York

Lecture Notes in Mathematics ISSN print edition: 0075-8434
                                ISSN electronic edition: 1617-9692

Library of Congress Control Number: 2011925676

Mathematics Subject Classification (2011): 60-02, 60G15, 55N35, 60Gxx, 60G60, 53Cxx, 62H35,
                                60G55, 53C65

Cover design: deblik, Berlin

Printed on acid-free paper

Springer is part of Springer Science+Business Media (www.springer.com)

Preface

Before you start reading them, we should tell you something about what you can expect to find in these lecture notes, and what you should not be looking for.

First and foremost, you should keep in mind that what we have here was written to be a companion for the Saint Flour Lectures, which cover twelve hours of lecture time in eight meetings. This is sufficient time to be able to give a good introduction to a subject, but it is not enough time to either teach it in depth or describe all its ramifications and applications. The notes reflect both the challenge and limitation inherent in these parameters.

The second point to keep in mind is that the title of the notes and the lectures is overly optimistic. When originally planning the lectures we had ambitious plans regarding the material that we hoped to cover. Eventually we managed to internalise the fact that there was only so much one could do in twelve hours and so the "smooth random functions" of the title are limited to Gaussian, and Gaussian-related (to be defined later), random functions. As you will see, "topological complexity" is also somewhat of a grandiose over-statement. However, perhaps by the time you finish reading the notes, especially Chap. 6, you will at least have a feeling for what we originally had in mind.

Also related to the structure of these notes is the fact that in 2007 we published a 450 page Springer monograph [8] *Random Fields and Geometry* (hereafter *RFG*) which more or less covers the theoretical aspects of these notes. Furthermore, we currently have on hand half of a second monograph [9] that, when completed, will cover a wide range of applications. This one will be called *Applications of Random Fields and Geometry: Foundations and Case Studies* (hereafter *ARFG*). Despite the tragic loss of our *ARFG* co-author Keith Worsley in February 2009, we plan still to complete this book, which will also be published by Springer. At the moment the earlier chapters, which include homework exercises for some of the material of these notes, can be found on our web pages.

Since this second volume will also probably grow to a size comparable to its theoretical precursor, one wonders if we really can have anything new or different to say in the 100 pages or so of these notes.

In some cases, we do. There have been developments of the theory since the first monograph came out, and some of these are touched on here. In particular, Sect. 4.10 describes a powerful, infinite dimensional, extension of the Gaussian kinematic formula worked out in detail in [82]. Chapter 5 will point you in the directions of applications, and will, hopefully, one day form the core of *ARFG*. Finally in Chap. 6 there is a brief discussion of some brand new results at the interface of random fields and *algebraic* topology. It is these topics that originally motivated the title of these notes, but, somehow, at Saint Flour there was not enough time to discuss them in any detail.

However, the main advantage of these notes is precisely that they are neither as exhaustive nor, we hope, as exhausting, as the two monographs.

Our main aim here then will be to give a readable and easily accessible introduction to an area that has been developing rapidly over the past few years, without getting bogged down with too much technical detail. For the missing details in the theory you can turn, for the most part, to *RFG*, and, for more details on applications, to *ARFG*.

Our secondary aim is to motivate you to download the existing chapters of *ARFG* and help us debug them, and, of course, to motivate you to order a copy of *RFG* from Springer.[1]

Finally, we have some acknowledgments to make. The first is to the Saint Flour scientific committee, for originally inviting RJA to give the course. RJA delivered most of the lectures, but JET also carried some of the load. Adding to this the fact that these notes are heavily based on our joint monograph *RFG*, the result is the current joint, yet again, product.

We are also grateful to the agencies that supported our research and writing during the last 2–3 years. In particular, JET thanks the National Science Foundation (DMS-0906801), RJA thanks the Israel Science Foundation (853/10) and both thank the Binational Science Foundation (2008262) and the NSF for a SGER grant (also with Shmuel Weinberger) that had a lot to do with getting Chap. 6 written.

Haifa, Israel   *Robert J. Adler*
Stanford, California   *Jonathan E. Taylor*

---

[1] By the way, the grandfather of both of these books was published in 1981 as *The Geometry of Random Fields* [2], and after being out of print for many years is newly available in the SIAM series *Classics in Applied Mathematics*. It is, of course, rather dated, but also rather readable, because in those days its author had not yet learnt how to make easy material look hard. Since then, he has.

Contents

Chapter 1
Introduction

Since you are a wise reader you certainly have, by now, read the Preface. So you already know, in general, what these notes are about. This chapter has some more details.

1.1 Random Fields on Stratified Manifolds

The "smooth random functions" of these notes are what we shall generally call "random fields", by which we mean random functions defined on parameter spaces M richer than the real line $\mathbb{R}$, where the parameter is generally taken to be 'time', t. We shall still use t to denote the parameter, along with Latin letters such as f and g to denote the random fields, but the typical M that we have in mind will be a compact domain of $\mathbb{R}^N$, $N > 1$, such as the cube $I_N \triangleq [0,1]^N$, or perhaps a manifold, such as the $(N-1)$-dimensional sphere of radius $\lambda > 0$, which we denote by $S_\lambda(\mathbb{R}^N)$. (We shall also write $S(\mathbb{R}^N)$ for $S_1(\mathbb{R}^N)$.)

The random function itself will generally take values in $\mathbb{R}^d$, $d \geq 1$, where there is typically no connection between the dimensions N and d.

In addition, we shall often add to this setup a deterministic function F, from $\mathbb{R}^d$ to $\mathbb{R}^{d'}$, which we shall use to define a new random field, $g : M \to \mathbb{R}^{d'}$, so that we have the structure of Fig. 1.1.1. Again, there need be no connection between d and d', although in most cases on interest d' will be one.

This general structure may at first seem a little strange. After all, why not define g directly without the intermediate function F? The gain, as we shall see later, is that very often it is possible to work, in this fashion, with quite simple f and F, but generating a rather complicated g, whose distributional properties would be difficult to derive directly.

The other thing that might strike you as a little strange is the shape of the set M in Fig. 1.1.1. After all, it is neither a simple cube nor a simple sphere. (In fact, it is really not even clear from the figure whether M is two or three dimensional; viz. whether or not it has a non-empty interior. For what

R.J. Adler and J.E. Taylor, *Topological Complexity of Smooth Random Functions*, Lecture Notes in Mathematics 2019, DOI 10.1007/978-3-642-19580-8_1, © Springer-Verlag Berlin Heidelberg 2011

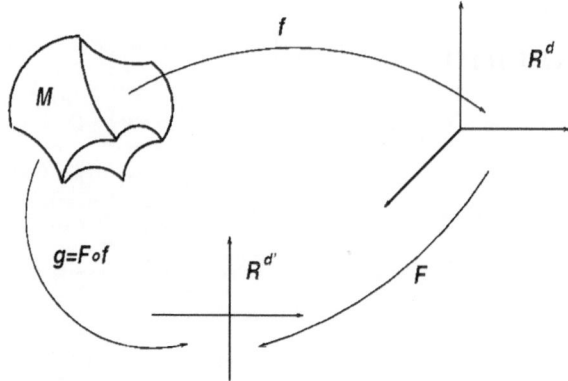

Fig. 1.1.1 A typical setting: $f : M \to \mathbb{R}^d$ random, $F : \mathbb{R}^d \to \mathbb{R}^{d'}$ deterministic, and $g = F \circ f : M \to \mathbb{R}^{d'}$

follows, however, we shall assume that M does have an interior, and so is 3-dimensional.)

In order to handle parameter spaces that are able to include both domains such as cubes, or simplicial complexes, with edges and corners, along with domains which are smooth manifolds or are bounded by unions of such manifolds, we shall generally work in the setting of *stratified manifolds*. These are basically sets that can be partitioned into the disjoint union of manifolds, so that we can write

$$M = \bigsqcup_{j=0}^{\dim M} \partial_j M, \tag{1.1.1}$$

where each stratum, $\partial_j M$, $0 \leq j \leq \dim(M)$, is itself a disjoint union of a number of j-dimensional manifolds. For the M of Fig. 1.1.1, $\partial_3 M$ is the interior of M; $\partial_2 M$ the collection of its two-dimensional sides, some concave and some convex; $\partial_1 M$ is made up of the one-dimensional edges; and $\partial_0 M$ contains the corner vertices. In each case, boundaries are not included. Thus $\partial_2 M$ includes neither the bounding edges, which are in $\partial_1 M$, nor the vertices, which are in $\partial_0 M$.

Later we shall have more to say about regularity conditions for stratified manifolds, but for the moment you will lose little if you restrict yourself to the three special cases: $M = I_N$, $M = S(\mathbb{R}^N)$, and $M = B(\mathbb{R}^N)$, where $B_\lambda(\mathbb{R}^N)$ is the ball of radius λ in $\mathbb{R}^N$ and $B(\mathbb{R}^N) \stackrel{\Delta}{=} B_1(\mathbb{R}^N)$.

In fact, when we get around to proving things, you might want to assume that M is a smooth manifold without boundary. This will make complicated notation and long summations collapse into simple single term expressions. For the main proof of the notes M will be a submanifold of a sphere.

1.2 Topological Complexity

Topology is the study of 'pure' shape, as opposed to geometry, which might be considered as adding considerations of size to those of shape. We shall be concerned with both geometry and topology. As far as geometry is concerned, the concepts that we shall require come from convex and integral geometry. Topology itself splits into many sub-disciplines, the main two of which, from our point of view, are differential topology and algebraic topology. Differential topology has a lot to do with concepts such as curvature, which are local objects, whereas algebraic topology tends to concentrate on global structures, such as homology and homotopy groups.

A point at which all of these subjects meet is in studying the Euler, or Euler–Poincaré, characteristic of an object. Euler characteristics will play an important rôle in these notes, so it is worthwhile to take a little time to meet them in a few of their guises.

Perhaps the most natural (and historically appropriate) way to start is to look at the ring of convex sets in $\mathbb{R}^N$, $\mathcal{C}^N$, and to search for an integer valued functional with the following two basic properties:

$$\varphi(M) = \begin{cases} 0 & \text{if } M = \emptyset, \\ 1 & \text{if } M \neq \emptyset \text{ is convex,} \end{cases} \tag{1.2.1}$$

and

$$\varphi(M_1 \cup M_2) = \varphi(M_1) + \varphi(M_2) - \varphi(M_1 \cap M_2), \tag{1.2.2}$$

whenever $M, M_1, M_2 \in \mathcal{C}^N$. That is, φ is a finitely additive functional that assigns the value one to convex sets.

An important result of integral geometry states that not only does a functional possessing these two properties exist, but it is *uniquely* determined by them. This functional is the Euler characteristic. One can do many computations from this definition alone, particularly if one is prepared to accept the fact that such a functional can be defined on far more general classes of sets, with the second part of (1.2.1) replaced by the requirement that $\varphi(M) = 1$ if M is homotopic to a ball.

To go a little further, suppose that M is a simplicial complex, or a triangulation of a stratified manifold.[1] Then an alternative, but equivalent, definition of the Euler characteristic of M is as the alternating sum

---

[1] That is, we have a covering of M by diffeomorphic images of simplices of dimension no more than $\dim(M)$, such that, if two such images have a non-empty intersection, then the pre-images of the intersection must be full facets (sub-simplices) of each of the original simplices. We shall abuse notation by writing M for both the manifold and the triangulation.

$$\varphi(M) = \sum_{j=0}^{\dim M} (-1)^j \alpha_j(M), \qquad (1.2.3)$$

where $\alpha_j(M)$ is the number of j-dimensional facets in the simplicial complex M or, respectively, its triangulation.[2]

While this definition of the Euler characteristic is algebraic, there is another, based on Morse theory, which is quite different. For simplicity, suppose that M is a C^2 manifold (without boundary) in $\mathbb{R}^N$. Take an $f \in C^2(\mathbb{R}^N)$ and consider its critical points, i.e. those points t satisfying $\nabla f(t) = 0$. Then, under mild regularity conditions on f that we shall meet later, another way to define the Euler characteristic of M is to set

$$\varphi(M) = \sum_{j=0}^{\dim M} (-1)^j n_j(M, f), \qquad (1.2.4)$$

where $n_j(M, f)$ is the number of critical points at which the Hessian of f has j negative eigenvalues.

The representations (1.2.3) and (1.2.4) at first seem to have little in common, other than they both (not coincidentally) involve an alternating sum. As there was an arbitrariness in (1.2.3) coming from the choice of triangulation, so there is an arbitrariness in (1.2.4), this time coming from the choice of f. Other differences between the two results are superficial. In particular, whereas, for simplicity, we have written (1.2.4) only for manifolds without boundary, it has a natural extension to stratified manifolds, in which the generalisations of the n_j also count certain types of critical points of f restricted to the $\partial_k M$ (cf. Sect. 3.8).

Yet another definition of the Euler characteristic of a set comes from differential geometry, and is based on notion of local curvature. The key result here is the Gauss–Bonnet, or Chern–Gauss–Bonnet Theorem. It has a long and impressive history, starting in the early nineteenth century with simple Euclidean domains. Names were added to the result as the setting became more and more general. Here is a simple version, which holds in the setting of compact, orientable, C^2 Riemannian manifolds (M, g) of dimension N, with Riemannian curvature tensor R and volume form Vol_g:

$$\varphi(M) = \begin{cases} \frac{(-1)^{N/2}}{(2\pi)^{N/2} N!} \int_M \mathrm{Trace}(R^{N/2}) \, \mathrm{Vol}_g & N \text{ even} \\ 0 & N \text{ odd.} \end{cases} \qquad (1.2.5)$$

---

[2] Despite the fact that there is no uniqueness for triangulations and so the right hand side of (1.2.3) would seem to depend on the triangulation, it is a basic theory of algebraic topology that the Euler characteristic is well defined and independent of the triangulation.

Again, there is a version for stratified manifolds. More importantly, however, is the fact that this definition of the Euler characteristic is qualitatively different to the two we have seen so far. There is no longer any alternating sum, and what is clearly a global qeometric quantity is now expressed in terms of purely local properties of the manifold. Even more striking is the fact that the whereas the right hand side of (1.2.5) depends of the Riemannian metric g through both the curvature tensor and the volume form, the left hand side is a purely topological measure that depends on neither of these. In fact, Gauss was so impressed by this fact that he called his original version of the result his *Theorema Egregium* (Remarkable Theorem), a description which is still appropriate today.

There is also a representation of the Euler characteristic that has it roots in convex geometry, growing out of a result known as Steiner's formula [57,71] and leading, via the Weyl tube formula to an entire class of geometric quantifiers which we shall call Lipschitz-Killing curvatures. To describe Steiner's formula, we first require the notion of a *tube*, of radius $\rho > 0$, around a set $M \subset \mathbb{R}^N$. This is easily defined as

$$\text{Tube}(M, \rho) = \{ t \in \mathbb{R}^N : d(t, M) \le \rho \} \tag{1.2.6}$$

where $d(t, M) \stackrel{\Delta}{=} \inf_{s \in M} |t - s|$ is the usual Euclidean distance from the point t to the set M.

Now let λ_N denote Lebesgue measure in $\mathbb{R}^N$. Then Steiner established the existence of numbers $\mathcal{L}_0(M), \ldots, \mathcal{L}_{\dim M}(M)$, such that, for convex M,

$$\lambda_N\left(\text{Tube}(M, \rho)\right) = \sum_{j=0}^{\dim M} \omega_{N-j} \rho^{N-j} \mathcal{L}_j(M), \tag{1.2.7}$$

where

$$\omega_j = \lambda_j(B(0,1)) = \frac{\pi^{j/2}}{\Gamma(\frac{j}{2} + 1)} \tag{1.2.8}$$

is the volume of the unit ball in $\mathbb{R}^j$. Weyl and others extended this result to far more general sets M, although then (1.2.7) holds only for $\rho < \rho_c$, where the critical radius ρ_c depends both on the local convexity of M as well as some global structure.

The numbers $\mathcal{L}_j(M)$ are known as the Lipschitz-Killing curvatures[3] of M, and $\mathcal{L}_0(M)$ is just another notation for the Euler characteristic of M.

---

[3] To some they may be more familiar as versions of the *Minkowski functionals*, which are given by

$$\mathcal{M}_j(A) = (j! \, \omega_j) \, \mathcal{L}_{N-j}(A). \tag{1.2.9}$$

We shall have much more to say about Lipschitz-Killing curvatures later on, but, for the moment, you might like to use (1.2.7) to check that, for $B_\lambda(\mathbb{R}^N)$, the N-dimensional ball of radius λ,

$$\mathcal{L}_j\left(B_\lambda\left(\mathbb{R}^N\right)\right) = \lambda^j \binom{N}{j} \frac{\omega_N}{\omega_{N-j}}, \tag{1.2.10}$$

for all $0 \leq j \leq N$, while for the sphere $S_\lambda(\mathbb{R}^N)$,

$$\mathcal{L}_j\left(S_\lambda\left(\mathbb{R}^N\right)\right) = \lambda^j \binom{N}{j} \frac{2\omega_N}{\omega_{N-j}}, \tag{1.2.11}$$

if $N - 1 - j$ is even, and 0 otherwise.

A slightly less tidy, but similarly simple, calculation gives that, for N-dimensional rectangles,

$$\mathcal{L}_j\left(\prod_1^N [0, T_j]\right) = \sum_{j_1 \ldots j_k} T_{j_1} \cdots T_{j_k}, \tag{1.2.12}$$

where the sum is over the $\binom{N}{k}$ *distinct* choices of k indices between 1 and N.

Note that, with the exception of the Euler characteristic $\mathcal{L}_0(M)$, all of the Lipschitz-Killing curvatures are dependent on the size of the set. In the examples above it is clear that they scale nicely, in the sense that

$$\mathcal{L}_j(\lambda M) = \lambda^j \mathcal{L}_j(M), \tag{1.2.13}$$

for all $\lambda > 0$ and all j, where $\lambda M = \{x : x = \lambda s \text{ for some } s \in M\}$. This scaling holds in general, and so the Lipschitz-Killing curvatures cannot be topological invariants[4] of M. In some sense, $\mathcal{L}_j(M)$ is a measure of the average j-dimensional cross-section of M, a fact that we shall formalise later via Crofton's Formula, (3.7.1).

The Lipschitz-Killing curvatures also appear in another fashion, in an extremely important theorem due initially to Hadwiger [47] in a setting a little more general than the convex ring and named after him. It states that if ψ is a real valued function on nice classes of sets in $\mathbb{R}^N$, invariant under rigid motions, additive (in the sense of (1.2.2)) and either continuous or monotone,[5]

Written in terms of the Minkowski functionals, (1.2.7) takes on the form of a classic Taylor series expansion. What is amazing about it is that the expansion is finite.

[4] In fact, as we shall see, there are analogues of the Lipschitz-Killing curvatures for sets M embedded in Riemannian manifolds, in which case, with the single exception of $\mathcal{L}_0(M)$, the $\mathcal{L}_j(M)$ all depend on the Riemannian metric.

[5] Monotonicity is in the sense that, for all pairs M_1, M_2, either $M_1 \subseteq M_2 \Rightarrow \psi(M_1) \leq \psi(M_2)$ or $M_1 \subseteq M_2 \Rightarrow \psi(M_1) \geq \psi(M_2)$. Although this seems a rather standard

then

$$\psi(M) = \sum_{j=0}^{N} c_j \mathcal{L}_j(M), \tag{1.2.14}$$

where $c_0, \ldots, c_N$ are (ψ-dependent) constants. Thus, studying intrinsic volumes is equivalent to studying a far wider class of functionals on sets. We shall apply this result on a number of occasions.

For a final definition of the Euler characteristic we move to a purely algebraic interpretation. Thus, suppose that $H_0, \ldots, H_{\dim M}$ are the homology groups of M, and that the dimension of H_k is the Betti number β_k. Then

$$\varphi(M) = \sum_{j=0}^{\dim M} (-1)^j \beta_j. \tag{1.2.15}$$

This is an approach that we shall turn to only towards the end of these notes.

Although we have now seen four quite different definitions of Euler characteristics, the truth is that we have only uncovered the tip of a large iceberg of approaches, motivated by different classes of sets M, different branches of geometry or topology, and different applications. Nevertheless, hopefully we have begun to convince you that the Euler characteristic is a topological characteristic that appears naturally in a wide range of settings, and, as such, provides a good starting measure of topological complexity.

That this is indeed the case, and that after working with Euler characteristics one sees natural extensions of basic results to far more sophisticated measures of complexity, is a fact that you will hopefully appreciate by the time you reach the end of the notes.

1.3 Random Fields and Complexity

There are many ways that one could think of for studying the topological complexity of random fields, but clearly what all have in common is that one needs to concentrate on aspects of the sample paths of the fields which are

---

requirement for a measure (and so for $\mathcal{L}_N$) note that the individual $\mathcal{L}_j$, $j < N$ are not themselves monotone.

Continuity is also a rather delicate issue. Essentially, the metric taken between sets is of the form

$$d(M_1, M_2) = \sum_{j=0}^{N} a_j \int_{\text{Graff}(N,N-j)} |\mathcal{L}_0(M_1 \cap V) - \mathcal{L}_0(M_2 \cap V)| \, dV,$$

for some constants a_j, where, for notation, see Sect. 3.7.

not only intrinsically interesting but also mathematically tractable from both topological and probabilistic viewpoints.

A natural place to start would be to look at the image of the parameter space M, or subsets of it, under the mapping $f : M \to \mathbb{R}^d$. For example, if M is a smooth manifold then the same will be true of $f(M)$. In general, it would be natural to describe the structure of $f(M)$ in terms of the topological structure of M, the probabilistic structure of f, and the dimensions $\dim(M)$ and d. Despite the fact that this seems like such an obvious question, we are not aware of any general results in this direction.

An inverse problem would be to look at sets in the parameter space M over which the random field exhibits specific behaviour. This is something for which *are* aware of many results, and describing them will form much of the core of these notes. These objects are of intrinsic interest, partly because they have many applications, and partly because they are analytically tractable and lead to an elegant theory which has implications beyond the original setting. We shall call the sets behind this theory *excursion sets* and define them as

$$A_D \; \equiv \; A(f, M, D) \; \overset{\Delta}{=} \; \{t \in M : f(t) \in D\} \; \equiv \; M \cap f^{-1}(D), \quad (1.3.1)$$

where $D \subset \mathbb{R}^d$. When $d = 1$ and D is a positive half line, excursion sets are often referred to as *nodal domains* or *super-level sets*, and we write, with some ambiguity,

$$A_u \; \equiv \; A_u(f, M) \; \overset{\Delta}{=} \; \{t \in M : f(t) \geq u\} \; \equiv \; A(f, M, [u, \infty)). \quad (1.3.2)$$

There is a central theorem in these notes and, one way or another, everything else is related to it. We call it the *Gaussian kinematic formula*, for reasons that we shall explain in a moment. While special cases appeared as early as the mid 1970s, the elegant geometric form in which it can be found in these notes first appeared in Taylor's 2001 McGill PhD thesis [79] and the paper [80]. It has grown considerably in content and importance since then, as we understand more and more of its structure and its applications. The statement, for $f = (f_1, \ldots, f_d) : M \to \mathbb{R}^d$, with component fields f_j which are independent, smooth, zero mean, unit variance, and Gaussian, and where M and $D \subset \mathbb{R}^d$ are nice enough, is

$$\mathbb{E}\left\{\mathcal{L}_i\left(A(f, M, D)\right)\right\} = \sum_{j=0}^{\dim M - i} \begin{bmatrix} i + j \\ j \end{bmatrix} (2\pi)^{-j/2} \mathcal{L}_{i+j}(M) \, \mathcal{M}_j^\gamma(D). \quad (1.3.3)$$

There is much here that needs explanation. The combinatorial coefficients $\begin{bmatrix} n \\ m \end{bmatrix}$ are known as *flag coefficients* and are defined by

$$\begin{bmatrix} n \\ k \end{bmatrix} = \frac{[n]!\,\omega_n}{[k]!\,[n-k]!\,\omega_n\omega_{n-k}}, \quad [n]! = n!\omega_n. \tag{1.3.4}$$

The $\mathcal{M}_j^\gamma$, described and defined in Sect. 3.5 below, are known as *Gaussian Minkowski functionals*. These, to a large extent, play the rôle of Lipschitz-Killing curvatures in Gauss space and turn out to be rather important quantities in their own right.[6]

The Lipschitz-Killing curvatures on both sides of (1.3.3) are computed with respect to a specific Riemannian metric induced on M by the random field f, in a way which we have yet to describe. Recall, however, that $\mathcal{L}_0(A)$ is the Euler–Poincaré characteristic, and so, when $j = 0$ in (1.3.3), $\mathcal{L}_0\left(A(f,M,D)\right)$ is independent of any Riemannian structure.

The result (1.3.3) has a long history. If M is a simple interval $[0,T]$, f is real valued *and stationary*, and $D = \{u\}$, then (1.3.3) is essentially the famous Rice formula, which gives the mean number of crossings of the level u by f, and dates back to 1939 [69] and 1945 [70]. This states that

$$\mathbb{E}\left\{\#\left\{t \in [0,T] : f(t) = u\right\}\right\} = T\frac{\lambda_2^{1/2}}{\pi}e^{-u^2/2}.$$

where $\lambda_2 \overset{\Delta}{=} \mathbb{E}\{\dot{f}^2(t)\}$.

Since the establishment of the original Rice formula there have been tens, if not hundreds, of papers extending it in many ways. Much of this work is summarised in the recent monograph of Azais and Wschebor [11], which is closer to the style of the original theory than are these notes.

In a different vein, back in the 1970s Adler took a more geometric approach, and proved an early version of the Gaussian kinematic formula relying only on integral geometry and some Morse theory. Furthermore, the sets M were restricted to being N-dimensional rectangles and f was univariate and stationary. The theory up until 1980 was summarised in [2].

During the 1990s a series of papers by the late Keith Worsley and coworkers appeared (e.g. [87, 89–91]) that were important precursors to the general theory of these notes. However, as already mentioned, it was in 2001 that the modern theory began.

So much for a brief history. We have a long and interesting path to treck before all the parts of (1.3.3) will make sense to us, let alone until we can prove (a special case of) it. Before doing so, here are a few reasons why, apart from mere intellectual curiousity, this treck might be worthwhile.

---

[6] In fact, they also have interesting infinite dimensional counterpart, described briefly in Sect. 4.10 where (in honour of the fact that these notes cover lectures given in France) we shall see how they can be expressed in terms of the operators of the Malliavin calculus. This is not something that can be done for the Euclidean LKCs.

1.3.1 Statistical Implications

The general structure of (1.3.3) has significant implications for a class of statistical problems out of the purely Gaussian scenario. As in Sect. 1.1 (cf. Fig. 1.1.1) taking $F : \mathbb{R}^k \to \mathbb{R}$ to be nice, define the (typically non-Gaussian) process

$$f(t) = F(g(t)) = F(g_1(t), \ldots, g_k(t)), \tag{1.3.5}$$

with g Gaussian as for (1.3.3). Then it follows immediately from (1.3.3) that

$$\mathbb{E} \left\{ \mathcal{L}_i \left(A(f, M, [u, \infty)) \right) \right\}$$
$$= \sum_{j=0}^{\dim M - i} \begin{bmatrix} i + j \\ j \end{bmatrix} (2\pi)^{-j/2} \mathcal{L}_{i+j}(M) \, \mathcal{M}_j^\gamma (F^{-1}[u, +\infty)). \tag{1.3.6}$$

Non-Gaussian processes of the form (1.3.5) appear naturally in a wide variety of statistical applications of smooth random fields (e.g. [4,8,9,74,87, 89,90] with an excellent introductory review in [88]).

An additional, and extremely important, application of (1.3.3) lies in the so called 'Euler characteristic heuristic' that, for a wide range of random fields f,

$$\left| \mathbb{P} \left\{ \sup_{t \in M} f(t) \geq u \right\} - \mathbb{E} \left\{ \mathcal{L}_0 \left(A(f, M, [u, \infty)) \right) \right\} \right| \leq error(u), \tag{1.3.7}$$

where $error(u)$ is of a *substantially* smaller order than both of the other terms as $u \to \infty$. In the Gaussian case, this heuristic is now a well established theorem, and the error term is known to be of order $\exp(-u^2(1 + \eta)/2)$ for some (often identifiable) $\eta > 0$, while both the probability and expectation are of order $u^{\dim M - 1} \exp(-u^2/2)$ [81]. The ability to compute the expectation therefore provides useful, explicit, approximations for the excursion probability, which is not explicitly computable expect for a handful of very special cases.

We shall have a lot more to say about the exceedence probability (1.3.7) later, specifically in Sect. 2.4 and in Chap. 5 when we turn to applications.

1.3.2 Connections with Sample Path Behaviour

One of the main reasons these notes center around the Gaussian kinematic formula is, to put it simply, that the formula exists.

This is no mean feat in an area in which closed form expressions are few and far between. Consider, for example, the following random variables, all defined for a real valued random field f:

- The number of connected components of the excursion set $A(f, M, [u, \infty))$.
- The number of local maxima of f above a level u.
- The number of critical points of f above a level u.
- The exceedence probability $\mathbb{P}\{\sup_M f(t) \geq u\}$.

For none of these do we have, nor are we ever likely to have, closed form expressions, even under restrictive assumptions such as stationarity and isotropy.

On the other hand, let $\psi(u)$ denote the expectation of any of the three random variables listed above, or the exceedence probability. Then we shall see later that, in every case, and in wide generality, it is true that

$$|\mathbb{E}\left\{\mathcal{L}_0\left(A(f, M, [u, \infty)))\right)\right\} - \psi(u)| \; \to \; 0, \quad \text{as } u \to \infty. \qquad (1.3.8)$$

In each case it is also possible to say something about the rate of convergence, which almost always turns out to be of the same type as that described above for (1.3.7).

The detailed, rigorous proofs of results such as (1.3.8) are purely analytic, and rely on representing $\psi(u)$ as a (usually high dimensional) integral coming from a general Kac-Rice formula that we shall call an 'expectation meta-theorem' and shall meet in Sect. 4.1.

However, there has to be more to these results than mere analytic coincidence, and the asymptotic equivalence of all these expectations must also have sample path explanations. We shall discuss these briefly in Sect. 5.1.

These equivalences, however, highlight the centrality and importance of the Gaussian kinematic formula. There are many random variables arising in the study of random fields, many of which are related. But there is only one family – the Lipschitz-Killing curvatures of excursion sets – which are fully amenable (at least in terms of their expectations) to mathematical analysis.

1.3.3 Geometry

One of the basic results of integral and convex geometry is the so-called kinematic fundamental formula which, in its simplest form, states that for nice subsets M_1 and M_2 of $\mathbb{R}^n$,

$$\int_{G_n} \mathcal{L}_i\left(M_1 \cap g_n M_2\right) d\nu_n(g_n) = \sum_{j=0}^{n-i} \begin{bmatrix} i+j \\ i \end{bmatrix} \begin{bmatrix} n \\ j \end{bmatrix}^{-1} \mathcal{L}_{i+j}(M_1)\mathcal{L}_{n-j}(M_2).$$

$$(1.3.9)$$

Here G_n is the isometry group of $\mathbb{R}^n$ with Haar measure ν_n normalised so that, for any $x \in \mathbb{R}^n$ and any Borel $A \subset \mathbb{R}^n$, $\nu_n\left(\{g_n \in G_n : g_n x \in A\}\right) = \mathcal{H}_n(A)$, where $\mathcal{H}_n$ is n-dimensional Hausdorff measure. (See [51, 73] for M_j elements of the convex ring or similar, and [16] for more esoteric M_j closer to the spirit of these notes.)

Now reconsider (1.3.3). Taking $(\Omega, \mathcal{F}, \mathbb{P})$ as the probability space on which f lives, (1.3.3) can be rewritten as

$$\int_\Omega \mathcal{L}_i\left(M \cap (f(\omega))^{-1}(D)\right) d\mathbb{P}(\omega) = \sum_{j=0}^{\dim M - i} \begin{bmatrix} i+j \\ j \end{bmatrix} (2\pi)^{-j/2} \mathcal{L}_{i+j}(M)\, \mathcal{M}_j^\gamma(D).$$

$$(1.3.10)$$

Written this way, it is clear on comparing (1.3.9) and (1.3.10) that the Gaussian kinematic formula can be interpreted as a kinematic formula over Gaussian function space, rather than over the isometry group on Euclidean space. From this observation comes the name of (1.3.10) as the Gaussian kinematic formula. The source of this 'coincidence', and its importance to geometry, will slowly become clear as you progress through the notes.

Chapter 2
Gaussian Processes

The theory of Gaussian processes and fields is rich and varied, and many excellent books have been written on the subject, among them Bogachev [15], Dudley [33], Fernique [39], Hida and Hitsuda [49], Janson [52], Ledoux and Talagrand [60], Lifshits [61] and Piterbarg [68], not to mention *RFG* and another old favourite of ours, another set of lecture notes, [3]. In particular, a new book [11] by Jean-Marc Azaïs and Mario Wschebor has recently appeared that has a lot of material similar, but generally complementary, to what interests us.

We have no intention to go into any detail in the current notes, however, and so will take a quick route towards defining Gaussian processes on general parameter spaces that will get us where we need to go with the minimum of fuss. All you will need to know to follow this is some rather basic graduate level probability, and the definition of the multivariate Gaussian distribution.[1] We shall start, however, with a very simple example which requires nothing beyond undergraduate probability and some innovative calculus, but which is already extremely instructive.

---

[1] Recall that a $\mathbb{R}^d$ valued random variable $X = (X_1, \ldots, X_d)$ is said to be *multivariate Gaussian* if, for every $\alpha = (\alpha_1, \ldots, \alpha_d) \in \mathbb{R}^d$, the real valued variable $\langle \alpha, X \rangle = \sum_{i=1}^{d} \alpha_i X_i$ is univariate Gaussian. In this case there exists a mean vector $m \in \mathbb{R}^d$ with $m_j = \mathbb{E}\{X_j\}$ and a non-negative definite $d \times d$ covariance matrix C, with elements $c_{ij} = \mathbb{E}\{(X_i - m_i)(X_j - m_j)\}$, such that the probability density of X is given by

$$\phi_d(x) = \frac{1}{(2\pi)^{d/2}|C|^{1/2}} e^{-(x-m)C^{-1}(x-m)'/2}, \qquad x \in \mathbb{R}^d, \tag{2.0.1}$$

where $|C| = \det(C)$. We write this as $X \sim N(m, C)$, or $X \sim N_d(m, C)$ if we need to emphasise the dimension, and also adopt the standard but heavily overworked symbol ϕ to denote the density ϕ_1 of a $N(0, 1)$ random variable.

R.J. Adler and J.E. Taylor, *Topological Complexity of Smooth Random Functions*, Lecture Notes in Mathematics 2019,
DOI 10.1007/978-3-642-19580-8_2, © Springer-Verlag Berlin Heidelberg 2011

2.1   The Cosine Process

Perhaps the grandfather of all smooth stochastic processes is the *cosine random process* on $\mathbb{R}$. It is defined as

$$f(t) \overset{\Delta}{=} \xi \cos \lambda t + \xi' \sin \lambda t, \qquad (2.1.1)$$

where ξ and ξ' are uncorrelated, equidistributed, random variables and λ is a positive constant.

It is elementary trigonometry to see that the cosine process can also be written as

$$f(t) = R \cos(\lambda(t - \theta)), \qquad (2.1.2)$$

where $R^2 = \xi^2 + (\xi')^2 \geq 0$ and $\theta = \arctan(\xi/\xi') \in (-\pi, \pi]$, from whence the name 'cosine process'. Assuming, for convenience, that $\mathbb{E}\{\xi\} = 0$, we have that the covariance function of f is given by

$$\begin{aligned}
C(s,t) &= \mathbb{E}\{f(s)f(t)\} \\
&= \mathbb{E}\{(\xi \cos \lambda s + \xi' \sin \lambda s)(\xi \cos \lambda t + \xi' \sin \lambda t)\} \\
&= \mathbb{E}\{\xi^2\} \cos(\lambda(t - s)),
\end{aligned}$$

on using the fact that ξ and ξ' are uncorrelated and equidistributed. Consequently, regardless of the distribution of ξ, the cosine process is stationary. (See Sect. 2.6 below for definitions of stationarity and isotropy.)

One of the nice aspects of the cosine process is that many things that are either difficult or impossible to compute for more general processes can be computed exactly, and from first principles, once some assumptions are made on the distribution of ξ. We shall therefore now assume that ξ and ξ' are independent, Gaussian variables, with zero mean and common variance σ^2. As an example of what can be computed, consider, for $u > 0$, the exceedence probability

$$\mathbb{P}\left\{ \sup_{0 \leq t \leq T} f(t) \geq u \right\}, \qquad (2.1.3)$$

which we met in the Introduction.

Under the Gaussian assumption, R^2 has an exponential distribution with mean $1/(2\sigma^2)$, θ has a uniform distribution on $(-\pi, \pi]$, and R and θ are independent. We can use this information to compute some exceedence probabilities directly, and start by defining the *number of upcrossings* by f of the level u in time $[0, T]$,

$$N_u = N_u(f, T) = \#\{t \in [0, T] : f(t) = u \text{ and } df(t)/d(t) > 0\}.$$

It is trivial to see that the exceedence probability that we are after can now be written as

$$\mathbb{P}\left\{\sup_{0\leq t\leq T} f(t) \geq u\right\} = \mathbb{P}\left\{f(0) \geq u\right\} + \mathbb{P}\left\{f(0) < u, N_u \geq 1\right\}$$

$$= \Psi\left(\frac{u}{\sigma}\right) + \mathbb{P}\left\{f(0) < u, N_u \geq 1\right\}. \qquad (2.1.4)$$

where

$$\Psi(x) \triangleq 1 - \Phi(x) \triangleq \frac{1}{\sqrt{2\pi}} \int_{-\infty}^{x} e^{-u^2/2}\, du. \qquad (2.1.5)$$

is the tail probability function for a standard Gaussian variable.

We now restrict attention to the case $T \leq \pi/\lambda$, in which case, since f has period $2\pi/\lambda$, the event $\{f(0) \geq u, N_u \geq 1\}$ is empty, implying that

$$\mathbb{P}\left\{f(0) < u, N_u \geq 1\right\} = \mathbb{P}\left\{N_u \geq 1\right\}.$$

Again using the fact that $T < \pi/\lambda$, note that N_u is either 0 or 1. In order that it be 1, two independent events must occur. Firstly, we must have $R > u$, with probability $e^{-u^2/2\sigma^2}$. Secondly (draw a picture) θ must fall in an interval of length λT, so that the final result is

$$\mathbb{P}\left\{\sup_{0\leq t\leq T} f(t) \geq u\right\} = \Psi\left(\frac{u}{\sigma}\right) + \frac{\lambda T}{2\pi\sigma} e^{-u^2/2\sigma^2}, \qquad (2.1.6)$$

and the probability density of the supremum is given by

$$\frac{1}{\sigma}\phi\left(\frac{u}{\sigma}\right) + \frac{\lambda T u}{2\pi\sigma^2} e^{-u^2/2\sigma^2}. \qquad (2.1.7)$$

This computation was so simple, that one is tempted to believe that it must be easy to extend to many other processes. In fact, this is not the case, and the cosine process and field, which we shall meet in a moment, are the *only* differentiable, stationary, Gaussian processes for which the exceedence probabilities are explicitly known.

However, before we leave it, we can use the cosine process to motivate a more general approach. Note first that since, as noted above, N_u is either 0 or 1 when $T < \pi/\lambda$, we can rewrite (2.1.4) as

$$\mathbb{P}\left\{\sup_{0\leq t\leq T} f(t) \geq u\right\} = \Psi\left(\frac{u}{\sigma}\right) + \mathbb{E}\{N_u\}. \qquad (2.1.8)$$

Thus, rather than arguing as above, we could concentrate on finding an expression for the mean number of upcrossings.

More importantly, note that for any T, and, indeed, for *any differentiable random process*, the above argument always gives

$$\mathbb{P}\left\{\sup_{0 \leq t \leq T} f(t) \geq u\right\} \leq \mathbb{P}\left\{f(0) \geq u\right\} + \mathbb{E}\{N_u\}. \tag{2.1.9}$$

Thus there would seem to be a close relationship between exceedence probabilities and level crossing rates, that actually becomes exact for the cosine process over certain intervals. In fact, since, for a one dimensional set, its Euler characteristic is given by the number of its connected components, the expectation in the right hand sides of both (2.1.8) and (2.1.9) could be written as $\mathbb{E}\{\varphi(A_u(f,T))\}$, where φ is the Euler characteristic.

2.2 The Cosine Field

The cosine field is a straightforward extension to $\mathbb{R}^N$ of the cosine process, and has the representation

$$f(t) = f(t_1, \ldots, t_N) \overset{\Delta}{=} \frac{1}{\sqrt{N}} \sum_{k=1}^{N} f_k(\lambda_k t_k), \tag{2.2.1}$$

where each f_k is the process on $\mathbb{R}$ given by

$$f_k(t) = \xi_k \cos t + \xi_k' \sin t.$$

The λ_k are fixed, and the ξ_k and ξ_k' are taken to be identically distributed and uncorrelated.

Again, it is a simple exercise to check that the cosine field is both stationary and isotropic but it is somewhat harder to compute its exceedence probabilities. To see what can be done, we restrict attention to the cosine process on a rectangle of the form $T = \prod_{k=1}^{N}[0, T_k]$. Then, given the structure of the cosine field as a sum, it is immediate that

$$\sup_{t \in T} f(t) = \frac{1}{\sqrt{N}} \sum_{k=1}^{N} \sup_{0 \leq t_k \leq T_k} f_k(t).$$

If we assume that the ξ_k and ξ_k' are all independent $N(0, \sigma^2)$, then the suprema of the individual f_k are also independent. Further assuming that each $T_k \in (0, \pi/\lambda_k]$, (2.1.6) and (2.1.7) give their individual distributions. The distribution of the supremum of the cosine field is then the convolution of these. The computations involved in actually doing the convolution are not easy, but Piterbarg [68] showed that, if $p_N(u)$ is the density function of the supremum, ϕ the standard Gaussian density and $\phi^{(k)}$ its k-derivative, then there are simple constants, C_{nk}, depending only on n and k, such that

$$p_N(u) = \phi\left(\frac{u}{\sigma}\right) + \sum_{k=1}^{N}(-1)^k C_{nk}\phi^{(k)}\left(\frac{u}{\sigma}\right)\sum_{j_1\ldots j_k}\prod_{i=1}^{k}\frac{\lambda_{j_i}T_{j_i}}{\sigma}. \qquad (2.2.2)$$

The inner sum here is over the $\binom{N}{k}$ subsets of size k of $\{1,\ldots,N\}$.

Now assume that all the λ_j are identical. Then, appropriately rewritten, this result will recall the Gaussian kinematic formula. Setting $\sigma^2 = 1$ for convenience, and recalling the definition of the Lipschitz-Killing curvatures of rectangles at (1.2.12), we can write

$$p_N(u) = \sum_{k=0}^{N}(-1)^k C'_{nk}\,\phi^{(k)}(u)\,\lambda^k \mathcal{L}_j(T). \qquad (2.2.3)$$

Going a little further, integrating over u, and applying some non-trivial asymptotics (cf. Sect. 2.5 of [68]) one finds that

$$\mathbb{P}\left\{\sup_{t\in T} f(t) \geq u\right\} = e^{-u^2/2}\sum_{k=0}^{N}C''_{nk}H_{k-1}(u)\lambda^k \mathcal{L}_k(T) + o\left(e^{-(1+\eta)u^2/2}\right),$$
$$(2.2.4)$$

for some $\eta > 0$, The Hermite polynomials H_n are defined by

$$H_n(x) = n!\sum_{j=0}^{\lfloor n/2\rfloor}\frac{(-1)^j x^{n-2j}}{j!\,(n-2j)!\,2^j}, \qquad n \geq 0,\ x \in \mathbb{R}. \qquad (2.2.5)$$

where $\lfloor a\rfloor$ is the largest integer less than or equal to a and

$$H_{-1}(x) \overset{\Delta}{=} \sqrt{2\pi}e^{x^2/2}\Psi(x), \qquad (2.2.6)$$

where Ψ is the tail probability (2.1.5) of a standard normal.

The easily checked fact that

$$\frac{d^j}{dx^j}e^{-x^2/2} = (-1)^j H_j(x)e^{-x^2/2}, \qquad (2.2.7)$$

along with (2.2.3) explains why Hermite polynomials arise in the exceedence probabilities of the cosine field.

In fact, it turns out Hermite polynomials will arise in expressions for exceedence probabilities of all real valued C^2 Gaussian fields. Furthermore, along with the factor $e^{-u^2/2}$, they can be written in terms of Gaussian Minkowski functionals, a fact that we shall prove in Sect. 3.5. Thus the main term in the right hand side of (2.2.4) is now very reminiscent of the right hand side of the Gaussian kinematic formula.

However, even as a stand-alone result, it is already fascinating in that it links exceedence probabilities to the geometry of the parameter space.

2.3 Constructing Gaussian Processes

Since the construction of cosine processes and fields as a sum of deterministic functions with random amplitudes worked so well, we now try something similar in general. Thus, with M a potential parameter space, choose a finite or infinite set of functions $\varphi_1, \varphi_2, \ldots, \varphi_j : M \to \mathbb{R}$ satisfying only

$$\sum_j \varphi_j^2(t) \ < \ \infty, \qquad \text{for all } t \in M. \tag{2.3.1}$$

Let $\xi_1, \xi_2, \ldots$ be a sequence of independent, mean zero, variance 1, Gaussian random variables, and define the random field $f : M \to \mathbb{R}$ by

$$f(t) = \sum_j \xi_j \varphi_j(t). \tag{2.3.2}$$

That the sum converges in L^2, for each fixed $t \in M$, is a consequence of (2.3.1). How f behaves, as a function of t, is another issue, that we shall turn to later. Clearly, though, the smoother the φ_j are, the better behaved f will be.

The mean of f is zero, and its covariance function is given by

$$C(s, t) \ = \ \mathbb{E}\{f(s)f(t)\} = \sum_j \varphi_j(s)\varphi_j(t). \tag{2.3.3}$$

So we have seen how to go from a sum like (2.3.2) to a covariance function. Usually, however, Gaussian processes are defined by their covariance functions, rather than vice versa, so let's make a couple of calculations and then try to work backwards. Firstly, define a class of functions

$$S \ = \ \Big\{u : M \to \mathbb{R} : u(\cdot) = \sum_{i=1}^n a_i C(s_i, \cdot), \ a_i \text{ real}, \ s_i \in M, \ n \geq 1\Big\}. \tag{2.3.4}$$

Define an inner product on S by

$$(u, v)_H = \Big(\sum_{i=1}^n a_i C(s_i, \cdot), \ \sum_{j=1}^m b_j C(t_j, \cdot) \Big)_H$$

$$= \sum_{i=1}^{n} \sum_{j=1}^{m} a_i b_j C(s_i, t_j). \tag{2.3.5}$$

It is easy to check that if $u \in S$, then the following unusual property holds:

$$(u(\cdot), C(t, \cdot))_H = u(t). \tag{2.3.6}$$

This is known as the *reproducing kernel* property. The completion of S under this above inner product is known as the *reproducing kernel Hilbert space* (RKHS) of f, and all its elements also satisfy the reproducing property.

What is most interesting in this construction is that it also works in the other direction. That is, given a positive definite function C on a space M, one can define the completion of the space S of (2.3.4) under the inner product of (2.3.5), find a an orthonormal basis $\{\varphi_k\}$ for $H(C)$ and define the Gaussian process (2.3.2). This will have C as its covariance function. The RKHS is now associated with C rather than f, and is denoted by $H(C)$.

For further details see *RFG* (or virtually any of the other texts mentioned at the beginning of this chapter) where you will also find a proof of the following, harder and much deeper, result, which holds under the implicit assumption, assumed throughout these notes, that we are dealing only with separable random processes.[2]

Theorem 2.3.1. *Suppose that C is a bounded, positive definite function, continuous on $M \times M$, and that*

$$\sup_{s,t \in M} \left| C(s,s) + C(t,t) - 2C(s,t) \right| < \infty. \tag{2.3.7}$$

Let f be defined from C as above. Then f is a.s. continuous, if, and only if, the sum (2.3.2) converges uniformly on M, with probability one.

For the French among you, here is an (almost familiar) example. The *Brownian sheet* is the zero mean, Gaussian, random field on the positive orthant $[0, \infty)^N$ with covariance function

$$\mathbb{E}\{W(s)W(t)\} = (s_1 \wedge t_1) \times \cdots \times (s_N \wedge t_N). \tag{2.3.8}$$

Replacing each j in the above sums by a multi-index $j = (j_1, \ldots, j_N)$, it is then not too hard to check that the φ_j for W are given, for W restricted

---

[2] Recall that a real valued random process is called separable if there exists a countable dense subset D of M and a fixed event N with $\mathbb{P}\{N\} = 0$ such that, for any closed $B \subset \mathbb{R}$ and open $I \subset T$,

$$\{\omega : f(t, \omega) \in B, \, \forall t \in I\} \, \Delta \, \{\omega : f(t, \omega) \in B, \, \forall t \in I \cap D\} \, \subset \, N,$$

where Δ denotes the usual symmetric difference operator.

to $[0,1]^N$, by

$$\varphi_j(t) = 2^{N/2} \prod_{i=1}^{N} \frac{2}{(2j_i + 1)\pi} \sin\left(\tfrac{1}{2}(2j_i + 1)\pi t_i\right).$$

When $N = 1$, W is the completely familiar Brownian motion. The corresponding expansion is due to Lévy, and the corresponding RKHS is known as Cameron–Martin space.

The message of this section should, by now, be clear. When dealing with continuous Gaussian processes, we lose no generality whatsoever by treating them as sums of deterministic functions with independent Gaussian coefficients. This will be important throughout these notes.

2.4 The Canonical Process on $S(\mathbb{R}^l)$

There is a school of thought that takes the basic ideas of the previous section even further. Note that, for any $t \in M$, the sequence $\widetilde{\varphi}(t) = \{\varphi_1(t), \varphi_2(t), \dots\}$ belongs to ℓ^2. (cf. (2.3.1).) Consider the image of M in ℓ^2 under the mapping $t \to x = \widetilde{\varphi}(t)$, denote it by B, and define a new Gaussian process $\widetilde{f}$ by setting

$$\widetilde{f}(x) = f\left(\widetilde{\varphi}^{-1}(x)\right), \tag{2.4.1}$$

assuming always that φ is one to one.[3] Note that

$$
\begin{aligned}
\mathbb{E}\left\{\widetilde{f}(x)\widetilde{f}(y)\right\} &= \mathbb{E}\left\{f\left(\widetilde{\varphi}^{-1}(x)\right) f\left(\widetilde{\varphi}^{-1}(y)\right)\right\} \\
&= \sum_j \varphi_j\left(\widetilde{\varphi}^{-1}(x)\right) \varphi_j\left(\widetilde{\varphi}^{-1}(y)\right) \\
&= \sum_j x_j y_j \\
&= \langle x, y \rangle_{\ell^2}.
\end{aligned}
\tag{2.4.2}
$$

---

[3] This is actually a perfectly reasonable assumption. If there are two different points $s, t \in M$ mapping to the same point in $S(\mathbb{R}^l)$, then we must have

$$\mathbb{E}\left\{[f(t) - f(s)]^2\right\} = \sum_{1}^{\ell} [\varphi_j(t) - \varphi_j(s)]^2 = 0,$$

which implies that $f(t)$ and $f(s)$ are, almost surely, identical, and so one of the points s, t can be dropped from the parameter set.

In other words, there is really only *one* Gaussian process. It is defined on a subset of ℓ^2 and its covariance function is the natural inner product on ℓ^2. It is known as the *isonormal process*, and all of its properties must be properties only of the parameter set B, and so accessible via the techniques of Banach spaces.

While we shall not exactly adopt this approach, and, to some extent, it would fail us if we did, it will be particularly helpful in certain special cases.

In particular, suppose that f has constant variance, which for notational simplicity we take to be one, and, somewhat more restrictively, that the expansion (2.3.2) is finite. Consequently,

$$f(t) = \sum_{j=1}^{l} \xi_j \varphi_j(t), \tag{2.4.3}$$

for some $1 \leq l < \infty$ and

$$\sum_{j-1}^{k} \varphi_j^2(t) = \mathbb{E}\left\{f^2(t)\right\} = 1. \tag{2.4.4}$$

Thus, the set $B = \widetilde{\varphi}(M)$ of the previous section is now embedded in $S(\mathbb{R}^l)$, the unit sphere of $\mathbb{R}^l$, and the random field defined on it can be easily extended to the entire sphere. The corresponding field is known as *canonical (isotropic) process on $S(\mathbb{R}^l)$*. It has covariance $C(s,t) = \langle s, t \rangle$, and can be realised as

$$\widetilde{f}(t) = \sum_{j=1}^{\ell} t_j \xi_j. \tag{2.4.5}$$

The isotropy comes from the fact that $C(s,t)$ is function of only the (geodesic) distance between s and t. (cf. Sect. 2.6 for a definition and discussion of isotropy.)

The diagram of Fig. 1.1.1 can now be modified somewhat. In fact, if we take d independent copies of f and $\widetilde{f}$ so that now $f = (f_1, \ldots, f_d)$ and $\widetilde{f} = (\widetilde{f}_1, \ldots, \widetilde{f}_d)$, we can write

$$f(t) = \widetilde{f}\left(\widetilde{\varphi}(t)\right) = \left(\widetilde{f} \circ \widetilde{\varphi}\right)(t).$$

The picture is now as in Fig. 2.4.1, where we have neglected the final mapping F in Fig. 1.1.1.

It turns out that for many purposes it suffices to work with the second half of the figure, from $\widetilde{\varphi} \to \mathbb{R}^d$. In the following two subsections we shall see two examples of this.

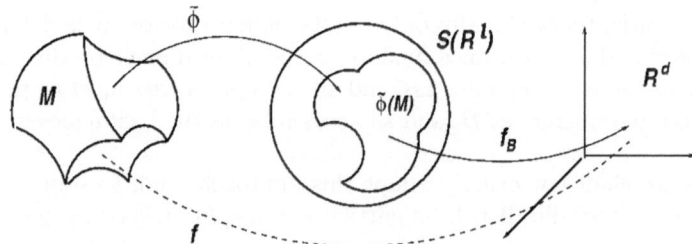

Fig. 2.4.1 The new setting with the canonical process on $S(\mathbb{R}^\ell)$ intervening

2.4.1 The Canonical Processes and Exceedence Probabilities

We are now going to look at the exceedence probabilities and continue the process of connecting them to geometry. The underlying technique is known as the *tube method* and has its roots in a pair of papers by Hotelling [50] and Weyl [86] in 1939. In the setting in which we shall apply it, it was developed primarily in [53, 58, 77].

Retaining the notation of the previous section, it is trivial that

$$\sup_{t \in M} f(t) \equiv \sup_{x \in \widetilde{\varphi}(M)} \widetilde{f}(x), \qquad (2.4.6)$$

so that in computing exceedence probabilities for unit variance, finite expansion Gaussian fields, we can concentrate first on treating only the canonical process over subsets of $S(\mathbb{R}^l)$. Thus, for the moment, let f be the canonical process on $S(\mathbb{R}^l)$, and let $M \in S(\mathbb{R}^l)$ be a nice set. Adopting the representation (2.4.5), we write $f(t)$ as $\langle \xi, t \rangle$, for $\xi \sim N(0, I_{\ell \times \ell})$ and $t \in S(\mathbb{R}^l)$.

Then we can argue as follows, writing $\mathbb{P}_{|\xi|}$ for the distribution of $|\xi|$:

$$
\begin{aligned}
\mathbb{P}\left\{ \sup_{t \in M} f_t \geq u \right\} &= \int_0^\infty \mathbb{P}\left\{ \sup_{t \in M} f_t \geq u \;\Big|\; |\xi| = r \right\} \mathbb{P}_{|\xi|}(dr) \\
&= \int_0^\infty \mathbb{P}\left\{ \sup_{t \in M} \langle \xi, t \rangle \geq u \;\Big|\; |\xi| = r \right\} \mathbb{P}_{|\xi|}(dr) \\
&= \int_u^\infty \mathbb{P}\left\{ \sup_{t \in M} \langle \xi, t \rangle \geq u \;\Big|\; |\xi| = r \right\} \mathbb{P}_{|\xi|}(dr) \\
&= \int_u^\infty \mathbb{P}\left\{ \sup_{t \in M} \langle \xi/r, t \rangle \geq u/r \;\Big|\; |\xi| = r \right\} \mathbb{P}_{|\xi|}(dr).
\end{aligned}
\qquad (2.4.7)
$$

Consider the integrand here. Since ξ is multivariate Gaussian, it is standard fare that the vector $\xi/|\xi|$ is uniformly distributed on $S(\mathbb{R}^l)$, independently of $|\xi|$, which is distributed as the square root of a χ_l^2 random variable. If we now write η_l to denote the uniform measure over $S(\mathbb{R}^l)$, we can rewrite the integrand as a simple volume computation, once we take a moment to consider tubes on spheres.

Our definition (1.2.6) of tubes extends from the simple Euclidean setting to subsets of spheres by adopting the standard geodesic metric on $S(\mathbb{R}^l)$ given by

$$\tau(s,t) = \cos^{-1}\left(\langle s,t \rangle\right).$$

Thus the tube of radius ρ around a closed set $M \in S(\mathbb{R}^l)$ is given by

$$
\begin{aligned}
\mathrm{Tube}(M,\rho) &= \left\{ t \in S(\mathbb{R}^l) : \tau(t,M) \leq \rho \right\} \\
&= \left\{ t \in S(\mathbb{R}^l) : \exists\, s \in M \text{ such that } \langle s,t \rangle \geq \cos(\rho) \right\} \\
&= \left\{ t \in S(\mathbb{R}^l) : \sup_{s \in M} \langle s,t \rangle \geq \cos(\rho) \right\}.
\end{aligned}
\tag{2.4.8}
$$

With this behind us, we can now continue the development of (2.4.7) to obtain

$$
\mathbb{P}\left\{ \sup_{t \in M} f_t \geq u \right\} = \int_u^\infty \eta_l\left(\mathrm{Tube}(M, \cos^{-1}(u/r))\right) \mathbb{P}_{|\xi|}(dr)
\tag{2.4.9}
$$

Thus, the exceedence probability that we seek is weighted average of the volume of tubes around M of varying radii, and if we could compute

$$\eta_l\left(\mathrm{Tube}(M,\rho)\right)$$

for all $\rho \leq 1$ we would, basically, be done, since the averaging, over the square root of a χ_l^2 random variable is, in principle, straightforward.

This approach – almost – works.

Firstly, not surprisingly, there are analogues of Steiner's formula (1.2.7), now called a tube formula, for subsets of spheres, with the Lipschitz-Killing curvatures appearing in the Euclidean case replaced by their spherical counterparts. We shall treat these in some detail in Chap. 3.4.2, but, for the moment, let us write them as $\mathcal{L}_j^1(M)$ so that, assuming the existence of a tube formula, (2.4.9) becomes

$$
\begin{aligned}
\mathbb{P}\left\{ \sup_{t \in M} f_t \geq u \right\} &= \sum_{j=0}^{\dim M} C_{\ell j} \mathcal{L}_j^1(M) \int_u^\infty \left(\cos^{-1}(u/r)\right)^{\ell-j} \mathbb{P}_{|\xi|}(dr), \\
&= \sum_{j=0}^{\dim M} C_{\ell j} \mathcal{L}_j^1(M) G_{\ell j}(u)
\end{aligned}
\tag{2.4.10}
$$

for some identifiable constants $C_{\ell j}$ and functions $G_{\ell j}$. Note that the final expression here is starting to take on the form of the right hand side of the Gaussian kinematic formula.

Where this argument breaks down is that the tube formula only works for small enough ρ or, in our case, small enough r. If r is large in the integrand of (2.4.7) then the tube around M has radius close to $\pi/2$, and it becomes easy, and, indeed, typical, for the tube to intersect itself 'on the other side' of the sphere in which it is embedded. Once a self-intersection of this kind occurs, tube formulae are no longer valid.

One way around this, which we shall not adopt in these notes, is to note that since the problems arise only for large r, and these have small probability under $\mathbb{P}_{|\xi|}$, one can ignore the tail of the integral, in a u-dependent fashion, and estimate the error involved in doing so. Then, however, (2.4.10) becomes an approximation rather than an exact result. We prefer to use the Euler characteristic approximation of (1.3.7) and will justify it later. In most cases, the two approaches yield identical approximations (cf. [78].)

The second problem with approaching everything via the canonical process on the sphere is that most random fields do not live in the sphere, and although the mapping from $M \to \widetilde{\varphi}(M)$ is natural one, in the final analysis one would like to have answers that depend not on the structure of $\widetilde{\varphi}(M)$, but rather on the structure of M and the covariance structure of f. This is, in fact, not too hard to do, and we shall see later how to relate the $\mathcal{L}_j^1(\widetilde{\varphi}(M))$ to the $\mathcal{L}_j(M)$.

The final problem with this approach, however, is highly non-trivial: Not all random fields have orthogonal expansions with only a finite number of terms. In fact, this is the exception rather than the rule. For example, *no* isotropic random field on $\mathbb{R}^N$ has a finite expansion! Nevertheless, the isonormal process on the sphere turns out to be the key example for generating results for general processes, as we shall see later.

2.4.2   The Canonical Process and Geometry

Returning now to the original random field f on M, consider how the excursion sets of f relate to those of $\widetilde{f}$. That is, what is the relation between

$$A_D = \{t \in M : f(t) \in D\} \quad \text{and} \quad \widetilde{A}_D = \{x \in \widetilde{\varphi}(M) : \widetilde{f}(x) \in D\}?$$

The first thing to notice is that since $\widetilde{\varphi}$ is one-one (already assumed) and if we assume that it is C^2 or smoother (in fact, it will always be at least C^4 for us) then the fact that $\widetilde{\varphi}$ is a diffeomorphism implies that the Euler characteristics of A_D and $\widetilde{A}_D$ will be identical. Consequently,

$$\mathbb{E}\{\mathcal{L}_0(A_D)\} = \mathbb{E}\{\mathcal{L}_0(\widetilde{A}_D)\}, \tag{2.4.11}$$

so that if we can compute the expected Euler characteristics of excursion sets for the canonical process on spheres, then we can, at least in principle, compute them for all Gaussian random fields with finite expansions.

Of course, we shall still face the same two problems that we faced above. The answers will depend on the structure of $\widetilde{\varphi}(M)$, rather than on the structure of M and the covariance structure of f, and they will only hold for random fields with finite expansions.

Furthermore, it is not at all clear if one can extend (2.4.11) to Lipschitz-Killing curvatures other than the Euler characteristic. For example, it is certainly not true in general that

$$\mathcal{L}_N(A_D) \equiv \mathcal{H}_N(A_D) = \mathcal{H}_N(\widetilde{A}_D) \equiv \mathcal{L}_N(\widetilde{A}_D),$$

where $\mathcal{H}_N(A_D)$ is the Euclidean volume of A_D but $\mathcal{H}_N(\widetilde{A}_D)$ is the surface area of $\widetilde{A}_D$ as a subset of the sphere. That there is nevertheless a way to obtain the general Gaussian kinematic formula, which gives an expression for means of all the $\mathcal{L}_j(A_D)$ from a parallel result for the canonical process, is one of the mysteries that will be unravelled as you proceed through these notes.

2.5   The Basic Theory of Gaussian Fields

To make the lecture course for which these notes were prepared complete and self-contained, we would have needed another 24 h or so to give a mini-course on Gaussian processes. Thus, for example, if you look at *RFG* (and by now you should have ordered a personal copy from Springer) you will see that the first third of the book is devoted to this material.

Clearly this was not possible. On the other hand, we do need some results from the general theory, and some specific moment results, for later use, and so they are collected in the following sections, with no attempt to prove anything. Everything is proven in *RFG* in full detail.

In fact, if you are reading through these notes by yourself, and have an impatient nature, you can actually skip these sections for now, go directly to the geometry of Chap. 3, and return later, as needed.

We should really begin by actually defining real valued *Gaussian (random) fields* or *Gaussian (random) processes*, something which have not actually done yet, as being a random fields for which the (finite dimensional) distributions of $(f_{t_1}, \ldots, f_{t_n})$ are multivariate Gaussian for each $1 \leq n < \infty$ and each collection $(t_1, \ldots, t_n) \in M^n$.

Since multivariate Gaussian distributions are determined by means and covariances, it is immediate that Gaussian random fields are determined by their mean and covariance functions defined, respectively, by

$$m(t) = \mathbb{E}\{f(t)\} \tag{2.5.1}$$

and

$$C(s,t) = \mathbb{E}\left\{(f(s) - m(s))\,(f(t) - m(t))\right\}. \qquad (2.5.2)$$

In fact, this is one of the main reasons, beyond ubiquitous but not always justified appeals to the central limit theorem, that Gaussian processes are such popular and useful choices for models for random processes on general spaces.

2.5.1 Regularity for Gaussian Process

We have already spoken about continuous and differentiable fields, but have said nothing about conditions that ensure this. In the Gaussian case, everything is dependent on the size of the parameter space, which we shall measure via the canonical metric.[4]

The *canonical metric*, d, of a zero mean Gaussian field on a topological space M, is defined by setting

$$d(s,t) \triangleq \left[\mathbb{E}\left\{(f(s) - f(t))^2\right\}\right]^{\frac{1}{2}}, \qquad (2.5.3)$$

in a notation that will henceforth remain fixed.[5] A ball in this metric, of radius ε and centered at a point $t \in M$ is denoted by

$$B_d(t,\varepsilon) \triangleq \{s \in M : d(s,t) \leq \varepsilon\}. \qquad (2.5.4)$$

Assume that M is d-compact, in the sense that

$$\mathrm{diam}(M) \triangleq \sup_{s,t \in M} d(s,t) < \infty. \qquad (2.5.5)$$

Fix $\varepsilon > 0$ and let $N(M,d,\varepsilon) \equiv N(\varepsilon)$ denote the smallest number of d-balls of radius ε whose union covers M. Set

$$H(M,d,\varepsilon) \equiv H(\varepsilon) = \ln\left(N(\varepsilon)\right). \qquad (2.5.6)$$

Then N and H are called the (metric) *entropy* and *log-entropy* functions for M (or f).

---

[4] There is also a more powerful approach based on the notion of *majorising measures* which we shall not adopt. For information on this approach see *RFG* and the far more serious treatment in [60].

[5] Actually, d is only a pseudo-metric, since although it satisfies all the other demands of a metric, $d(s,t) = 0$ does not necessarily imply $s = t$.

Here then is the main result about Gaussian continuity and boundedness, due originally, more or less in the form given below, to Richard Dudley [31, 32]. It is not the latest word in the topic, but it will more than suffice for our purposes. Note how the topological and geometric structure of M blend together with the covariance structure of f to give a measure, the metric entropy, which determines everything in this result.

This blending of the geometry of the parameter space together with a metric derived from the random field will also lie at the heart of the Gaussian kinematic formula, although it will be different geometry and a different metric.

Theorem 2.5.1. *Let f be a centered Gaussian field on a d-compact M, d the canonical metric, and H the corresponding log-entropy. Then there exists a universal constant K such that*

$$\mathbb{E}\left\{\sup_{t\in M} f_t\right\} \leq K \int_0^{\text{diam}(M)} H^{1/2}(\varepsilon)\,d\varepsilon, \tag{2.5.7}$$

and

$$\mathbb{E}\left\{\omega_{f,d}(\delta)\right\} \leq K \int_0^{\delta} H^{1/2}(\varepsilon)\,d\varepsilon, \tag{2.5.8}$$

where

$$\omega_{f,d}(\delta) \stackrel{\triangle}{=} \sup_{d(s,t)\leq\delta} |f(t) - f(s)|, \quad \delta > 0, \tag{2.5.9}$$

Furthermore, there exists a random $\eta \in (0,\infty)$ and a universal constant K such that

$$\omega_{f,d}(\delta) \leq K \int_0^{\delta} H^{1/2}(\varepsilon)\,d\varepsilon,$$

for all $\delta < \eta$.

A complement to this result states that f is also stationary, then

f is a.s. continuous on M $\iff$ f is a.s. bounded on M

$$\iff \int_0^{\delta} H^{1/2}(\varepsilon)\,d\varepsilon < \infty, \quad \forall \delta > 0. \tag{2.5.10}$$

For necessary and sufficient conditions in the general case one needs to turn to the notion of majorising measures mentioned above.

2.5.2 *Gaussian Fields on* $\mathbb{R}^N$

The entropy conditions above yield very simple sufficient conditions for continuity of centered Gaussian fields on compact sets M of $\mathbb{R}^N$. In fact, it is easy to check that, defining

$$p^2(u) \overset{\Delta}{=} \sup_{|s-t|\leq u} \mathbb{E}\left\{|f_s - f_t|^2\right\}, \tag{2.5.11}$$

a.s. continuity and boundedness follow if, for some $\delta > 0$, either

$$\int_0^\delta (-\ln u)^{\frac{1}{2}}\, dp(u) < \infty \quad \text{or} \quad \int_\delta^\infty p\left(e^{-u^2}\right) du < \infty. \tag{2.5.12}$$

Furthermore, there exists a constant K', dependent only on the dimension N, and a random $\delta_o > 0$, such that, for all $\delta < \delta_o$,

$$\omega_f(\delta) \leq K' \int_0^{p(\delta)} (-\ln u)^{\frac{1}{2}}\, dp(u), \tag{2.5.13}$$

where the modulus of continuity ω_f is as in (2.5.9), but taken with respect to the usual Euclidean metric rather than the canonical one. A similar bound, in the spirit of (2.5.8), holds for $\mathbb{E}\{\omega_f(\delta)\}$.

A sufficient condition for either integral in (2.5.12) to be finite is that, for some $0 < K < \infty$ and $\alpha, \eta > 0$,

$$\mathbb{E}\left\{|f_s - f_t|^2\right\} \leq \frac{K}{|\log|s - t||^{1+\alpha}},$$

for all s, t with $|s - t| < \eta$. Related conditions hold on the spectral density in the stationary case. See *RFG* for details.

In practical situations, it is rare indeed that one even gets close to the logarithmic behavior of (2.5.14). The more common situation is that the covariance function has a power series representation of the form

$$C(s,t) = C(t,t) + (t - s)\Lambda_t(t - s)' + o\left(|t - s|^{2+\delta}\right), \tag{2.5.14}$$

for $|t - s|$ small and some $\delta > 0$, or, in the stationary case

$$C(t) = C(0) + t\Lambda t' + o\left(|t|^{2+\delta}\right), \tag{2.5.15}$$

for t in the neighborhood of the origin. The matrices Λ_t and Λ are $N \times N$ and positive definite.

2.5.3 Differentiability

Since we shall also be requiring that our random functions are a.s. C^2, a few words on this condition are also in order. Firstly, unlike continuity, which requires nothing of the parameter space M other than it have a topology (so one can talk about continuity) differentiability requires that M itself has a differentiable structure. For the moment, we limit ourselves to $\mathbb{R}^N$ with its usual structure.

It then turns out that, at least in the Gaussian scenario, differentiability can be handled within the framework of continuity since derivatives, if they exist, must still be Gaussian. Since this is an important observation, that has been missed by many authors in the past, we shall deviate from the policy of this section and actually give details of how to do things.

To start, we need to define L^2 derivatives. Choose a point $t \in \mathbb{R}^N$ and a sequence of k 'directions' $t'_1, \ldots, t'_k$ in $\mathbb{R}^N$, and write these as $t' = (t'_1, \ldots, t'_k)$. We say that f has a k-th order L^2 partial derivative at t, in the direction t', if the limit

$$D^k_{L^2} f(t, t') \stackrel{\Delta}{=} \lim_{h_1, \ldots, h_k \to 0} \frac{1}{\prod_{i=1}^{k} h_i} \Delta^k f(t, t', h) \qquad (2.5.16)$$

exists in mean square, where $h = (h_1, \ldots, h_k)$. Here $\Delta^k f(t, t', h)$ is the symmetrized difference

$$\Delta^k f(t, t', h) = \sum_{s \in \{0,1\}^k} (-1)^{k - \sum_{i=1}^{k} s_i} f\left(t + \sum_{i=1}^{k} s_i h_i t'_i\right)$$

and the limit in (2.5.16) is interpreted sequentially, i.e. first send h_1 to 0, then h_2, etc. Note that if f is Gaussian then so are its L^2 derivatives, when they exist.

By choosing $t' = (e_{i_1}, \ldots, e_{i_k})$, where e_i is the vector with i-th element 1 and all others zero, we can talk of the mean square partial derivatives

$$\frac{\partial^k}{\partial t_{i_1} \ldots \partial t_{i_k}} f(t) \stackrel{\Delta}{=} D^k_{L^2} f(t, (e_{i_1}, \ldots, e_{i_k})) \qquad (2.5.17)$$

of f of various orders.

Moving now to almost sure differentiability, first endow the space $\mathbb{R}^N \times \otimes^k \mathbb{R}^N$ with the norm

$$\|(s, s')\|_{N,k} \stackrel{\Delta}{=} |s| + \|s'\|_{\otimes^k \mathbb{R}^N} = |s| + \left(\sum_{i=1}^{k} |s'_i|^2\right)^{1/2},$$

and write $B_{N,k}(y,h)$ for the ball centered at $y = (t,t')$ and of radius h in the metric induced by $\| \cdot \|_{N,k}$. Furthermore, write

$$M_{k,\rho} \triangleq M \times \{t' : \|t'\|_{\otimes^k \mathbb{R}^N} \in (1-\rho, 1+\rho)\}$$

for the product of M with the ρ-tube around the unit sphere in $\otimes^k \mathbb{R}^N$. This is enough to allow us to formulate

Theorem 2.5.1. *Suppose f is a centered Gaussian random field on an open $M \in \mathbb{R}^N$, possessing k-th order partial derivatives in the L^2 sense in all directions everywhere inside M. Suppose, furthermore, that there exists $0 < K < \infty$, and $\rho, \delta, h_0 > 0$ such that for $0 < \eta_1, \eta_2, h < h_0$,*

$$\mathbb{E}\left\{\left[\eta_1^{-k}\Delta^k f(t,t',\eta_1 1) - \eta_2^{-k}\Delta^k f(s,s',\eta_2 1)\right]^2\right\} \tag{2.5.18}$$
$$< K \left| \ln\left(\|(t,t') - (s,s')\|_{N,k} + |\eta_1 - \eta_2|\right)\right|^{-(1+\delta)},$$

for all

$$((t,t'),(s,s')) \in M_{k,\rho} \times M_{k,\rho} : (s,s') \in B_{N,k}((t,t'),h),$$

where $\eta_j 1$ denotes the k-vector all of whose elements are η_j. Then, with probability one, f is k times continuously differentiable.

Proof. Recalling that we have assumed the existence of L^2 derivatives, we can define the Gaussian field

$$\widehat{f}(t,t',\eta) = \begin{cases} \Delta^k f(t,\eta t') & \eta \neq 0, \\ D_{L^2}^k f(t,t') & \eta = 0, \end{cases}$$

where $D_{L^2}^k f$ is the mean square derivative (2.5.16). This process is defined on the parameter space $\widehat{M} \triangleq M_{k,\rho} \times (-h,h)$, an open subset of the finite dimensional vector space $\mathbb{R}^N \times \otimes^k \mathbb{R}^N \times \mathbb{R}$, with norm

$$\|(t,t',\eta)\|_{N,k,1} = \|(t,t')\|_{N,k} + |\eta|.$$

Whether or not f is k times differentiable on M is clearly the same issue as whether or not $\widehat{f}$ is continuous in $\widehat{M}$, with the issue of the continuity of $\widehat{f}$ really being only on the hyperplane where $\eta = 0$. But this puts us back into the setting of the previous subsection, and it is easy to check that condition (2.5.14) there translates to (2.5.18) in the current scenario.   □

As for continuity, it is rare in practice to get close to the upper bound in (2.5.18), and this condition will easily be satisfied if, in analogy to (2.5.14) and (2.5.15), the covariance function has a Taylor series expansion of up to order $2k$ with a remainder of $o(|h|^{2k+\eta})$ for some $\eta > 0$.

2.6 Stationarity, Isotropy, and Constant Variance

Although we have already met both stationarity and isotropy more than once, the time has now come to define them properly and list some of their basic properties.

We start by noting that a random field on a general parameter space M is called (second order) *stationary*, or *homogeneous*, if it has constant means and the covariance function $C(s,t)$ is a function of the difference $s-t$ only. With some abuse of notation, we shall write $C(s,t) = C(s-t)$.

Of course, if M is general, there is no reason why $s,t \in M$ implies that $s-t$ is also in M, and so it is necessary to assume that M has a group structure. In these notes, when discussing stationarity, we shall be concerned only with the cases $M = \mathbb{R}^N$ or $M = S_\lambda(\mathbb{R}^N)$. Note that for Gaussian processes this definition of stationarity also implies what is known as *strong stationarity*, which is that the finite dimensional distributions of the field are invariant under translations.

A stationary field is called *isotropic* if the covariance function is direction independent, in the sense that $C(t) = C(|t|)$.

We now restrict attention to random fields on $\mathbb{R}^N$. There are two basic results in the theory of stationary processes. One is known as the *spectral distribution theorem* and one as the *spectral representation theorem*. The first, which is the only one that we shall need in these notes, is due originally to Bochner in a non-probabilistic setting. For fields on $\mathbb{R}^N$ it states that if a continuous function $C : \mathbb{R}^N \to \mathbb{R}$ is non-negative definite, and so the covariance function of a stationary random field, if and only if there exists a finite measure ν on the Borel σ-field $\mathbb{B}^N$ of $\mathbb{R}^N$ such that

$$C(t) = \int_{\mathbb{R}^N} e^{i\langle t,\lambda \rangle} \, \nu(d\lambda), \tag{2.6.1}$$

for all $t \in \mathbb{R}^N$.

The measure ν is called the *spectral measure* and, since C is real, must be symmetric, in the sense that $\nu(A) = \nu(-A)$ for all $A \in \mathbb{B}^N$. Similarly, if C is isotropic then ν must be spherically symmetric, in the sense that $\nu(A) = \nu(\Theta A)$ for all $A \in \mathbb{B}^N$ and any rotation Θ.

2.6.1 Spectral Moments and Derivatives of Random Fields

Given the spectral representation (2.6.1) we define the *spectral moments*

$$\lambda_{i_1 \dots i_N} \triangleq \int_{\mathbb{R}^N} \lambda_1^{i_1} \cdots \lambda_N^{i_N} \, \nu(d\lambda), \tag{2.6.2}$$

for all $(i_1, \ldots, i_N)$ with $i_j \geq 0$. Note that, since ν is symmetric, the odd ordered spectral moments, when they exist, are zero; i.e.

$$\lambda_{i_1 \ldots i_N} = 0, \qquad \text{if } \sum_{j=1}^{N} i_j \text{ is odd.} \qquad (2.6.3)$$

Spectral moments turn out to be closely related to the variances and covariances of derivatives of random fields.

Recalling the notion of mean square partial derivatives from (2.5.17) it is a straightforward exercise to check that, in general, their covariance functions are be given by

$$\mathbb{E}\left\{ \frac{\partial^k f(s)}{\partial s_{i_1} \partial s_{i_1} \ldots \partial s_{i_k}} \frac{\partial^k f(t)}{\partial t_{i_1} \partial t_{i_1} \ldots \partial t_{i_k}} \right\} = \frac{\partial^{2k} C(s,t)}{\partial s_{i_1} \partial t_{i_1} \ldots \partial s_{i_k} \partial t_{i_k}}. \qquad (2.6.4)$$

When f is stationary, the corresponding variances and covariances also have a nice representation in terms of spectral moments. For example, if f has mean square partial derivatives of orders $\alpha + \beta$ and $\gamma + \delta$ for $\alpha, \beta, \gamma, \delta \in \{0, 1, 2, \ldots\}$, then

$$\mathbb{E}\left\{ \frac{\partial^{\alpha+\beta} f(t)}{\partial^\alpha t_i \partial^\beta t_j} \frac{\partial^{\gamma+\delta} f(t)}{\partial^\gamma t_k \partial^\delta t_l} \right\} = (-1)^{\alpha+\beta} \frac{\partial^{\alpha+\beta+\gamma+\delta}}{\partial^\alpha t_i \partial^\beta t_j \partial^\gamma t_k \partial^\delta t_l} C(t)\bigg|_{t=0} \qquad (2.6.5)$$

$$= (-1)^{\alpha+\beta} i^{\alpha+\beta+\gamma+\delta} \int_{\mathbb{R}^N} \lambda_i^\alpha \lambda_j^\beta \lambda_k^\gamma \lambda_l^\delta \, \nu(d\lambda).$$

Note that although this equation seems to have some asymmetries in the powers, these disappear due to the fact that all odd ordered spectral moments, like all odd ordered derivatives of C, are identically zero.

Here are some important special cases of the above, for which we adopt the shorthand $f_j = \partial f / \partial t_j$ and $f_{ij} = \partial^2 f / \partial t_i \partial t_j$ along with a corresponding shorthand for the partial derivatives of C.

(a) f_j has covariance function $-C_{jj}$ and thus variance $\lambda_{2e_j} = -C_{jj}(0)$, where e_j is the vector with a 1 in the j-th position and zero elsewhere.

(b) In view of (2.6.3), and taking $\alpha = \gamma = \delta = 0$, $\beta = 1$ in (2.6.5)

$$f(t) \text{ and } f_j(t) \text{ are uncorrelated}, \qquad (2.6.6)$$

for all j and all t. If f is Gaussian, this is equivalent to independence. Note that (2.6.6) does *not* imply that f and f_j are uncorrelated *as processes*. In general, for $s \neq t$, we will have that $\mathbb{E}\{f(s)f_j(t)\} = -C_j(s - t) \neq 0$.

(c) Taking $\alpha = \gamma = \delta = 1$, $\beta = 0$ in (2.6.5) gives that

$$f_i(t) \text{ and } f_{kl}(t) \text{ are uncorrelated} \qquad (2.6.7)$$

for all i, k, l and all t. Again, if f is Gaussian, this is equivalent to independence.

Under the additional condition of isotropy, with its implication of spherical symmetry for the spectral measure, the structure of the spectral moments simplifies significantly, as do the correlations between various derivatives of f. In particular, it follows immediately from (2.6.5) that

$$\mathbb{E}\left\{f_i(t)f_j(t)\right\} = -\mathbb{E}\left\{f(t)f_{ij}(t)\right\} = \lambda_2 \delta_{ij} \qquad (2.6.8)$$

where δ_{ij} is the Kronecker delta and λ_2 is the *second spectral moment*

$$\lambda_2 \triangleq \int_{\mathbb{R}^N} \lambda_i^2 \, \nu(d\lambda), \qquad (2.6.9)$$

which, because of isotropy, is independent of the value of i. Consequently, if f is Gaussian, then the first order derivatives of f are independent of one another, in addition to being independent of f itself.

Finally, we note that a similar argument shows that even if f is neither stationary nor isotropic, but does have constant variance, then it is still true that f and its first order derivatives are uncorrelated.

2.6.2 Local Isotropy and the Induced Metric

Of all the relationships between spectral moments in the previous subsection, the most important is probably (2.6.8), which describes the lack of correlation between first order derivatives of random fields under isotropy. It turns out that, in the case of Gaussian fields, this makes many computations that are, a priori, quite forbidding actually quite easy. Thus it is not surprising that the theory of Gaussian fields began with the isotropic case.

It is not in general possible to transform non-isotropic fields to isotropic ones, but there are a number of ways to ensure that first order derivatives are uncorrelated. This property is important enough that we shall give it a name, defining random fields with constant mean and variance, and uncorrelated first order derivatives, to be *locally isotropic*.

It turns out that it is easy to transform non-isotropic but stationary random fields f on $\mathbb{R}^N$ to locally isotropic ones. If Λ is the $N \times N$ matrix of second spectral moments λ_{ij}, then it is trivial to check that the field defined by $\widetilde{f}(t) = f(\Lambda^{-1/2}t)$ is locally isotropic. (cf. (2.5.14).)

In the non-stationary case there is no such simple transformation available. However, there is a trick, based on Riemannian geometry, that allows one to change the Riemannian structure of the parameter space by introducing a Riemannian metric related to the covariance function that makes all first order *Riemannian* derivatives uncorrelated. It was this trick that, in many ways, was one of the most important themes of *RFG*, and is what allows one to move from a theory of stationary random fields on subsets of $\mathbb{R}^N$ to non-stationary fields on stratified manifolds. We shall see how this works later, in Sect. 4.5 when we introduce this special (induced) metric at (4.5.1).

2.7 Three Gaussian Facts

We close this chapter with three facts about multivariate Gaussian random variables that we shall need later. All are well known and easy to check, and we include them now only so that they will be easy to refer back to later.

It follows from the form (2.0.1) of the multivariate Gaussian density that if $X \sim N_d(m, C)$ then its characteristic function is given by

$$\phi(\theta) = \mathbb{E}\{e^{i\langle\theta,X\rangle}\} = e^{i\langle\theta,m\rangle - \theta C\theta'/2}, \tag{2.7.1}$$

where $\theta \in \mathbb{R}^d$. From this follows the fact that, if A is a $d \times d$ matrix, then

$$XA \sim N(mA, A'CA). \tag{2.7.2}$$

Next, if $n < d$, make the partitions

$$X = \left(X^1, X^2\right) = ((X_1, \ldots, X_n), \ (X_{n+1}, \ldots X_d)),$$
$$m = \left(m^1, m^2\right) = ((m_1, \ldots, m_n), \ (m_{n+1}, \ldots m_d)),$$
$$C = \begin{pmatrix} C_{11} & C_{12} \\ C_{21} & C_{22} \end{pmatrix},$$

where C_{11} is an $n \times n$ matrix. Then each X^i is $N(m^i, C_{ii})$ and the conditional distribution of X^i given X^j is also Gaussian, with mean vector

$$m_{i|j} = m^i + (X^j - m^j)C_{jj}^{-1}C_{ji} \tag{2.7.3}$$

and covariance matrix

$$C_{i|j} = C_{ii} - C_{ij}C_{jj}^{-1}C_{ji}. \tag{2.7.4}$$

Finally, we quote a fundamental moment result known as *Wick's formula*. This states that if $X = (X_1, X_2, \ldots, X_d) \sim N(0, C)$ then, for any non-negative integer m,

$$\mathbb{E}\left\{X_1 X_2 \cdots X_{2m+1}\right\} = 0, \tag{2.7.5}$$

$$\mathbb{E}\left\{X_1 X_2 \cdots X_{2m}\right\} = \sum \mathbb{E}\{X_{i_1} X_{i_2}\} \cdots \mathbb{E}\{X_{i_{2m-1}} X_{i_{2m}}\}$$

$$= \sum C(i_1, i_2) \cdots C(i_{2m-1}, i_{2m}), \tag{2.7.6}$$

where the sum is taken over the $(2m)!/m!\,2^m$ different ways of grouping $X_1, \ldots, X_{2m}$ into m pairs. Wick's formula can be proven by successive differentiation of the characteristic function (2.7.1).

Chapter 3
Some Geometry and Some Topology

In the preceeding two chapters we spent most of our time talking about probability, while making lots of tangential references to geometry and topology. The time has now come to be a little more formal in defining the concepts that we shall need from these disciplines.

Again, however, we remind you that these notes are meant to cover the spirit of 12 h of lectures, and so cannot be expected to be complete and definitive. As always, you can find a lot of the missing details in *RFG*, of which almost a third is devoted to setting things up properly. But, even there, the treatment is not complete. Both geometry and topology are subjects that require significant tomes of their own.

3.1 Some Notation for Riemannian Manifolds

We are going to assume that you are familiar with the basic notions of manifold theory. (For the French among you, we have been assured that manifolds are taught at about the same age at which French mathematicians are taught how to count.) If you are unfamiliar with manifold theory, think of M, in all that follows, as a sphere, torus, or all of $\mathbb{R}^N$. Better still, think of M as the solid ball or solid doughnut, which are manifolds with infinitely differentiable boundaries. In most applications it is manifolds with boundaries that are important. To be honest, in most applications the boundaries are only piecewise smooth, a fact that leads to the need to treat the stratified manifolds introduced in Sect. 3.3.

Now for the notation:

As usual, we write $T_t M$ for the tangent space to a manifold M at t. We shall typically denote elements of T_t by X_t, Y_t, etc, and denote differentiation by such vectors as $X_t f$. The collection of all tangent spaces forms the tangent bundle $T(M)$ of M.

A *Riemannian metric* g on M is a family, $\{g_t\}_{t \in M}$, of inner products on the tangent spaces $T_t M$.

R.J. Adler and J.E. Taylor, *Topological Complexity of Smooth Random Functions*, Lecture Notes in Mathematics 2019, DOI 10.1007/978-3-642-19580-8_3, © Springer-Verlag Berlin Heidelberg 2011

A C^k manifold M together with a C^{k-1} Riemannian metric g is called a C^k *Riemannian manifold* (M, g).

The (Riemannian) gradient of a C^1 function on a Riemannian manifold (M, g) is the unique continuous vector field on $T(M)$ such that

$$g_t(\nabla f_t, X_t) = X_t f \tag{3.1.1}$$

for every vector field X. If $M = \mathbb{R}^N$ with the usual Euclidean metric, then ∇f is the usual gradient. In general, however, ∇f 'knows' about the geometry of M, both 'physical' and 'Riemannian'.

Given two vector fields X and Y on M, with $X_t, Y_t \in T_t M$ for all $t \in M$, we use the usual notation $\nabla_X Y$ to denote the *covariant derivative* of Y in the direction X. Note that, unless M is flat, $\nabla_X Y(t)$ is quite different to the usual derivative $X_t Y_t$ of Y in the direction X. If M is a surface in $\mathbb{R}^N$, $\nabla_X Y$ is basically the projection of the latter onto the tangent spaces of M. Thus, like the gradient, $\nabla_X Y(t) \in T_t M$, and, like the gradient, it 'knows' about the geometry of M.

The last differential object that we require is the *(covariant) Hessian* $\nabla^2 f$ of a function $f \in C^2(M)$ defined either as the bilinear symmetric map from $C^1(T(M)) \times C^1(T(M))$ to $C^0(M)$ satisfying

$$\nabla^2 f(X, Y) \triangleq XYf - \nabla_X Yf = g(\nabla_X \nabla f, Y), \tag{3.1.2}$$

or, more simply, as

$$\nabla^2 f = \nabla(\nabla)f.$$

In the simple Euclidean case the Hessian is defined via the $N \times N$ matrix

$$H_f = (\partial^2 f / \partial x_i \partial x_j)_{i,j=1}^N,$$

so that $\nabla^2 f(X, Y) = X H_f Y'$.

We also need concepts of curvature. The first is the *Riemannian curvature operator*, R, which measures the failure of covariant derivatives to commute, and which is given by

$$R(X, Y) \triangleq \nabla_X \nabla_Y - \nabla_Y \nabla_X - \nabla_{[X,Y]}. \tag{3.1.3}$$

where the Lie bracket $[X, Y] = XY - YX$.

The *(Riemannian) curvature tensor*, also denoted by R, is defined by

$$\begin{aligned} R(X, Y, Z, W) &\triangleq g\left(\nabla_X \nabla_Y Z - \nabla_Y \nabla_X Z - \nabla_{[X,Y]} Z, \ W\right) \\ &= g(R(X, Y)Z, \ W), \end{aligned} \tag{3.1.4}$$

where the R in the second line is, obviously, the curvature operator. It is easy to check that for $\mathbb{R}^N$, equipped with the standard Euclidean metric, $R \equiv 0$, whereas on the sphere $S_\kappa(\mathbb{R}^N)$ we have $R \equiv -1/\kappa$.

Now suppose that M is embedded in some larger manifold $\widetilde{M}$. Then the *second fundamental form* of M in $\widetilde{M}$ is the operator S from $T(M) \times T(M)$ to $T^\perp(M)$, the collection of normal complements of the $T_t M$ in $T_t \widetilde{M}$, satisfying

$$S(X,Y) \overset{\Delta}{=} P^\perp_{T(M)} \left(\widetilde{\nabla}_X Y \right) = \widetilde{\nabla}_X Y - \nabla_X Y, \qquad (3.1.5)$$

where the equality here is known as *Gauss's formula*.

Now let ν denote a unit normal vector field on M, so that $\nu_t \in T_t^\perp M$ for all $t \in M$. Then the *scalar second fundamental form* of M in $\widehat{M}$ for ν is defined, for $X, Y \in T(M)$, by

$$S_\nu(X,Y) \overset{\Delta}{=} \widehat{g}\left(S(X,Y), \nu \right), \qquad (3.1.6)$$

where the internal S on the right hand side refers to the second fundamental form of (3.1.5). When there is no possibility of confusion we shall drop the qualifier 'scalar' and refer also to S_ν as the second fundamental form.

Finally, if M is orientable, which we shall henceforth assume without further comment, the metric g determines a volume form, which we shall write either as Vol_g or as $\mathcal{H}_{\dim M}$, the latter generally denoting Hausdorff measure based on the geodesic metric associated with the Riemannian metric g.

3.2 Coarea Formula

At the heart of the tube formula is another important result, due to Federer [37] and known as his *coarea formula*. The coarea formula allows us to break up integrals over manifolds into iterated integrals over submanifolds of lower dimension, and is a generalisation of the classical change of variables formula.

To state it, consider a differentiable map $f : M \to N$ between two Riemannian manifolds, with $m = \dim(M) \geq n = \dim(N)$, and define the generalised Jacobian

$$Jf(t) \overset{\Delta}{=} \sqrt{\det\left(g_t \left(\nabla f_i(t), \nabla f_j(t) \right) \right)}. \qquad (3.2.1)$$

The coarea formula states that for differentiable[1] f and for $g \in L^1(M, \mathcal{H}_m)$,

$$\int_M g(t) \, Jf(t) \, d\mathcal{H}_m(t) = \int_N d\mathcal{H}_n(u) \int_{f^{-1}(u)} g(s) \, d\mathcal{H}_{m-n}(s). \quad (3.2.2)$$

Simplifying matters a little, consider two special cases. If $M = \mathbb{R}^N$ and $N = \mathbb{R}$, both with the usual Euclidean metric, it is easy to see that $Jf = |\nabla f|$, so that

$$\int_{\mathbb{R}^N} g(t) \, |\nabla f(t)| \, dt = \int_{\mathbb{R}} du \int_{f^{-1}\{u\}} g(s) \, d\mathcal{H}_{N-1}(s), \quad (3.2.3)$$

where ∇f is the usual Euclidean gradient.

Another important special case arises when $M = N = \mathbb{R}^N$, so that $Jf = |\det \nabla f|$, and

$$\int_{\mathbb{R}^N} g(t) \, |\det \nabla f(t)| \, dt = \int_{\mathbb{R}^N} du \int_{f^{-1}\{u\}} g(s) \, d\mathcal{H}_0(s)$$

$$= \int_{\mathbb{R}^N} \left(\sum_{t:\, f(t)=u} g(t) \right) du. \quad (3.2.4)$$

Of course, if f is one-one, and the integrals are restricted to measurable domains, then this is no more than

$$\int_A g(u) \, du = \int_{f(A)} g(t) \, |\det \nabla f(t)| \, dt,$$

the usual change of variables formula.

We shall see that while (3.2.3) lies at the heart of the tube formula of the following section, (3.2.4) lies at the heart of the Kac-Rice metatheorem of Sect. 4.1.

3.3 Stratified Manifolds

Although there is a way to avoid it, unfortunately the natural setting in which to build a topological theory of excursion sets involves sets with corners and edges. There are two reasons for this. One is that in many applications the

---

[1] Federer's setting is actually somewhat more general than this, since it also holds for Lipschitz mappings. In this format derivatives are replaced with 'approximate derivatives' throughout.

parameter set of a random field itself has corners. The cube is the archtypical example. The other is that even if one works with parameter sets such as bounded domains with C^∞ boundaries, and with C^∞ fields f, even the simple excursion sets $A_u = M \cap \{t : f(t) \geq u\}$ will typically have corners at the points $t \in \partial M$ at which $f(t) = u$.

The only way around this problem is to work with smooth fields on parameter spaces that are manifolds without boundary, and, if you like, you can read the remainder of these notes believing that to be the case. It will make many formulae much simpler, but it will make them extremely limited as far as applications are concerned, with the possible expectation of $S(\mathbb{R}^N)$ for $N = 2, 3$.

There are many ways to build a theory of sets with corners, and, for reasons that should become clearer later, we shall take the path of stratified manifolds satisfying a collection of generally unrestrictive regularity conditions. The basic reference for these are the monographs by Goresky and Macpherson [43] and Pflaum [67]. However, as usual, you can find the details we need in *RFG*.

To define stratified manifolds, we start with a topological subspace M of a C^k ambient manifold $\widetilde{M}$. The term 'stratified' refers to a decomposition of M into strata which we take to be also be C^k manifolds. In particular, as in Sect. 1.1, we have the decomposition

$$M = \bigsqcup_{j=0}^{\dim M} \partial_j M \qquad (3.3.1)$$

where $\partial_j M$ is the *j-dimensional boundary* of M made up from the disjoint union of a finite number of j-dimensional manifolds. There is nothing unique in the decomposition (3.3.1).

The first regularity condition that we shall require relates to how the respective j-dimensional boundaries $\partial_j M$ are 'glued' together. The assumption that we shall make is that they are glued in such a way that M is a *Whitney stratified manifold* satisfying 'Condition B'.. Without going into details (see *RFG*) here are two examples:

1. The curve $M = \{(x, y) : x \in [0, \pi], \ y = \sin(x)\}$ is a Whitney stratified manifold satisfying Condition B. The boundary $\partial_0 M$ contains the two points $(0, 0)$ and $(\pi, 0)$, while $\partial_1 M$ is the remainder of the curve.
2. The curve $M = \{(x, y) : x \in [0, \pi], \ y = x \sin(1/x)\}$ is a stratified manifold with the same decomposition, but its behaviour near the origin precludes the regularity conditions.

The second of these examples is typical, in that one needs to be a little inventive to find non-regular examples.

We shall also assume that our stratified manifolds are C^2 and, for technical reasons related to the Morse theory we shall introduce later, are embedded in a manifold that is at least C^3. In addition, we shall require that they are cone spaces of arbitrary depth, C-tame for some finite C and locally convex.[2] All of these are defined in *RFG*, and all, bar the local convexity, are conditions meant to circumvent the imaginative examples that pure mathematicians are so fond of but that appear so rarely in applications. We shall call stratified manifolds satisfying all of these conditions *'regular'*.

Examples of regular stratified manifolds abound, and include:

- Piecewise linear sets.
- Finite simplicial complexes.
- Reimannian polyhedra.
- Reimannian manifolds (with boundary).
- Closed semialgebraic (subanalytic) subsets of Euclidean spaces. (i.e. sets which are finite Boolean combinations of excursion sets of algebraic (analytic) functions.)
- Elements of the convex ring with piecewise C^2 boundary.

Furthermore, regular stratified manifolds have many desirable properties, among them:

- They can be triangulated.
- They have a well defined Euler characteristic, definable as in (1.2.3) via the triangulation.
- The intersection of two regular stratified manifolds is generally a regular stratified space, with strata which are the intersections of the strata of the two manifolds.

3.4 Tube Formulae and Lipschitz-Killing Curvatures

Back in Sect. 1.2 we already met a simple version of the Weyl tube formula, which held for simple Euclidean sets and went by the name of Steiner's formula, cf. (1.2.7). We now want to develop a parallel result for N-dimensional, C^2, stratified manifolds M embedded in either $\mathbb{R}^l$ or $S_\lambda(\mathbb{R}^l)$, $l \geq N$. For example, in the first case, we want to show that, for sufficiently small $\rho \geq 0$, there exist numbers $\mathcal{L}_j(M)$, such that

$$\mathcal{H}_l\left(\text{Tube}(M, \rho)\right) = \sum_{i=0}^{N} \rho^{l-i} \omega_{l-i} \mathcal{L}_i(M), \qquad (3.4.1)$$

---

[2] A set is called *locally convex* if all its support cones are convex. Note that local convexity does not imply convexity. For example, the C^∞ manifolds $S(\mathbb{R}^N)$, considered as subsets of $\mathbb{R}^N$, are locally convex for all N, but they are definitely not convex. The shape ∨ has a convex support cone at every point other than at its base, but this suffices to stop it from being locally convex.

where $\mathcal{H}_l$ is l-dimensional Hausdorff measure. Furthermore, and importantly, we want to know how to calculate the $\mathcal{L}_j$. (We leave the analogous statement of the spherical case for later.)

3.4.1 Describing Tubes

For the moment there is little to gain (and, in view of what will come later, much to lose) by restricting ourselves to submanifolds of $\mathbb{R}^l$ and $S_\lambda(\mathbb{R}^l)$. Thus, until further notice, we shall let our manifolds M be quite general. However, in addition to assuming that M is embedded in an ambient manifold $\widetilde{M}$, we shall also need to allow for the fact that $\widetilde{M}$ itself may be embedded in a larger manifold $\widehat{M}$, to which we assign dimension l. (Think of $M = S^1 = S(\mathbb{R}^2)$), a one dimensional circle embedded in the two-dimensional plane $\widetilde{M} = \mathbb{R}^2$. If $\widehat{M} = \mathbb{R}^3$, then the tube around M, as defined by (3.4.2), is a solid torus.)

With the triple

$$M \subset \widetilde{M} \subset \widehat{M}$$

in mind, we establish the following convention about notation: In general, a $\widehat{\cdot}$ will denote an object when we are considering M as being embedded in $\widehat{M}$. For example, we write the normal cones of M as subsets of $\widetilde{M}$ and $\widehat{M}$ by $\widetilde{N_t M}$ and $\widehat{N_t M}$, respectively. Note for further reference that

$$\widehat{N_t M} = \widetilde{N_t M} \oplus T_t \widetilde{M}^\perp$$
$$= \left\{ X_t \in T_t \widehat{M} : X_t = V_t + Y_t, \ V_t \in \widetilde{N_t M}, \ Y_t \in T_t \widetilde{M}^\perp \right\},$$

where $T_t \widetilde{M}^\perp$ is the orthogonal complement of $T_t \widetilde{M}$ in $T_t \widehat{M}$.

Similarly, we denote the Riemannian metrics on $M, \widetilde{M}$ and $\widehat{M}$ by $g, \widetilde{g}$ and $\widehat{g}$, with similar notation for curvature tensors and second fundamental forms. Throughout we assume that the metrics are consistent, in the sense that the metrics on the smaller, embeddedm manifolds are pullbacks of those on the larger manifolds. We also have Riemannian (Hausdorff) volume on each of the manifolds, which we denote by $\mathcal{H}_n$ for the appropriate dimension n.

With the worst of the notation out of the way, we can now start looking at tubes. The tube about M in $\widehat{M}$, of radius ρ, is

$$\text{Tube}(M, \rho) = \left\{ x \in \widehat{M} : d_{\widehat{M}}(x, M) \le \rho \right\}, \tag{3.4.2}$$

where $d_{\widehat{M}}$ is geodesic distance on $\widehat{M}$. Examples, to which we shall often return, are given in Fig. 3.4.1. There M is either a one-dimensional circle embedded in the two-dimensional sphere $\widetilde{M} = S(\mathbb{R}^3)$ and the tube is the

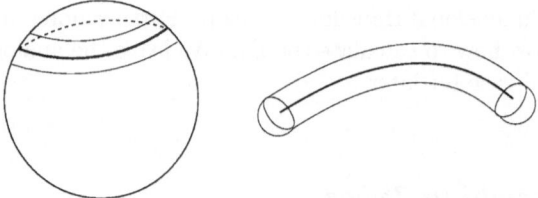

Fig. 3.4.1 Tubes around a circle embedded in a sphere and around a curve in $\mathbb{R}^3$

annulus enclosed between the highest and lowest circles, or M is a curve in space, in which case the tube is a bent cylinder with rounded ends.

With tubes defined, we can now begin, but shall not finish, a description of how to derive tube formulae. Our basic aim is to go as far as is necessary to show you how curvatures and second fundamental forms enter the picture, skimping on detail for most of the derivation, and then stopping exactly where the details start to involve long, problem specific, computations.

The first new concept that we shall need is that of the *critical radius* $\rho_c = \rho_c(M, \widehat{M})$ of a manifold M. In general, tube formulae are only valid for values of $\rho < \rho_c$. The first reason that we require ρ to be small is that we want to start with a linearization of the metric projection $\xi_M : \widehat{M} \to M$ given by

$$\xi_M(s) \overset{\Delta}{=} \operatorname*{argmin}_{t \in M} d_{\widehat{M}}(s,t), \tag{3.4.3}$$

which, for ρ small enough, parameterizes the tube as the *disjoint* union

$$\text{Tube}(M, \rho) = \bigcup_{t \in M} \left\{ s \in \widehat{M} : \xi_M(s) = t, \, d_{\widehat{M}}(s,t) \leq \rho \right\}. \tag{3.4.4}$$

Our first requirement on ρ_c is that it be less than the supremum of those ρ for which this union is, in fact, disjoint.

For example, for the embedded circle in Fig. 3.4.1, it is clear that ρ has to be small enough that the top circle – the upper boundary of the tube – does not reach the north pole. If it does, we lose the uniqueness required for ξ_M to be well defined and for there to be a natural way for (3.4.4) to hold.

Another way that things can go wrong is if M is sharply concave. For example, while for the smooth curve of Fig. 3.4.1 ρ_c is clearly positive, if we replace the curve by the shape $\vee$ we would have $\rho_c = 0$. This is why we earlier required that our stratified manifolds be locally convex; i.e. that their support cones[3] $\mathcal{S}_t M$ are convex for every $t \in M$. This will ensure that $\rho_c > 0$.

---

[3] The support cone of a point $t \in M$ is basically the closure of the cone of all vectors originating at t and which remain in M for (at least) an arbitrarily short distance.

There is another way to write the tube (3.4.4), which relies on local linearization and exponential maps. This second way will put us into a situation that is natural for applying the coarea formula for Sect. 3.2.

Note first that

$$\left\{ s \in \widehat{M} : \xi_M(s) = t,\ \widehat{d}(s,t) \le \rho \right\}$$
$$= \exp_{\widehat{M}} \left(\left\{ X_t \in T_t\widehat{M} : P_{T_t\widetilde{M}} X_t \in N_t M,\ |X_t| \le \rho \right\} \right).$$

Here $\exp_{\widehat{M}}$ is the exponential map from $T_t\widehat{M} \to \widehat{M}$ for generic $t \in \widehat{M}$, $P_{T_t\widetilde{M}}$ is projection from $T_t\widehat{M} \to \widehat{N_tM}$, and $|X_t|$ is the norm induced by $\widehat{g}$.

Therefore, for ρ small enough,

$$\text{Tube}(M,\rho) = \bigcup_{t \in M} \bigcup_{\{X_t \in T_t\widehat{M} : |X_t| \le \rho\}} \exp_{\widehat{M}}(t, X_t)$$
$$= \bigcup_{t \in M} \bigcup_{\{X_t \in \widehat{N_tM} : |X_t| \le \rho\}} \exp_{\widehat{M}}(t, X_t).$$

We shall also take ρ small enough for the second expression here to be a disjoint union. We continue to denote the largest such ρ by ρ_c.

Now we make the small step that is in fact the major leap in deriving tube formulae:

For $\rho < \rho_c$ it now follows from the above that the $\text{Tube}(M,\rho)$ is the image of the region

$$\left\{ (t, X_t, r) : t \in M,\ X_t \in \widehat{N_tM} \cap S(T_t\widehat{M}),\ 0 \le r \le \rho \right\} \qquad (3.4.5)$$

under the bijection $\widehat{F}$ defined by

$$\widehat{F}(t, X_t, r) \overset{\triangle}{=} \exp_{\widehat{M}}(t, rX_t). \qquad (3.4.6)$$

If you have trouble following all the symbols, look back to the curve in Fig. 3.4.1. Here both the exponential map and $\widehat{F}$ are identity mappings. In (3.4.5) the normal cone at each interior point is the plane perpendicular to the tangent line at the point. Letting X_t move over the 'sphere' in this plane traces out a circle, and allowing r to run between 0 and ρ fills out a disc of radius ρ. The union of these over t gives the cylindrical part of the tube. The two end regions can be handled similarly.

Now try it with the spherical example of Fig. 3.4.1, in which $\exp_{\widehat{M}}$ and $\widehat{F}$ are no longer the identity.

Why is all this important? Because we are now in a position to use the coarea formula.

3.4.2 Computing Volumes

Consider again the set in (3.4.5) and stratify it according to the stratification of M. In particular, let $D_j(\rho)$ be the stratum obtained by restricting t to the $\partial_j M$ there. The mapping (3.4.6) is then slightly different over each $D_j(\rho)$ and we acknowledge this by writing the restriction to $D_j(\rho)$ as F_j, $j = 0, \dots, \dim(M)$, dropping the hat off F.

With this notation, it is immediate that the volume of the tube around M is given by

$$\mathcal{H}_l(\mathrm{Tube}(M, \rho)) = \sum_{j=0}^{N} \mathcal{H}_l(F_j(D_j(\rho))), \qquad (3.4.7)$$

where we remind you that at each point a $\mathcal{H}$ appears we are using it to denote the Riemannian (or Hausdorff) volume on the appropriate space.

Now we can apply the coarea formula, to write each volume here as

$$\mathcal{H}_l(F_j(D_j(\rho))) = \int_0^\rho \int_{S_j(r)} \mathbb{1}_{F_j(D_j(\rho))}(x) \, \mathcal{H}_{j,r}(dx) dr, \qquad (3.4.8)$$

where

$$S_j(r) \triangleq \{s \in \widehat{M} : d_{\widehat{M}}(s, \partial_j M) = r\}, \qquad (3.4.9)$$

and $\mathcal{H}_{j,r}$ is the volume form induced on $S_j(r)$ by $\mathcal{H}_l$.

Alternatively, we can pull back $\mathcal{H}_{j,r}$ to the level sets making up $D_j(\rho)$ to obtain

$$\mathcal{H}_l(F_j(D_j(\rho))) = \int_0^\rho \int_{\partial_j M \times S(\mathbb{R}^{l-j})} \mathbb{1}_{D_j(\rho)}(t, s, r) \, F_{j,r}^*(\mathcal{H}_{j,r}) dr, \quad (3.4.10)$$

where $F_{j,r}$ is the partial map

$$F_{j,r}(t, s) \triangleq F_j(t, s, r)$$

and $F_{j,r}^*(\mathcal{H}_{j,r})$ is the pullback of $\mathcal{H}_{j,r}$.

In some sense, we are now done, and all that remains is some multivariate calculus, which, in the (in)famous words of Hermann Weyl [86], could be "accomplished by any student in a course of calculus". In fact, in order to turn (3.4.10) into a useable form, one needs only compute the pullback measures $F_{j,r}^*(\mathcal{H}_{j,r})$. This, in turn, requires computing some Jacobians and these, in turn, will involve the curvature and second fundamental form of M.

The actual computations are rather involved, and you can find details of them in *RFG* or any text on tubes. What it leads to is the definition of the following quantities, known as *Lipschitz-Killing curvature measures* (cf. [37])

which, for $\dim(\widetilde{M}) = \dim(\widehat{M}) = N = l$, and $0 \leq i \leq N$, are given by

$$\mathcal{L}_i(M, A) = \sum_{j=i}^{N} (2\pi)^{-(j-i)/2} \sum_{m=0}^{\lfloor \frac{i-i}{2} \rfloor} C(N-j, j-i-2m) \frac{(-1)^m}{m!\,(j-i-2m)!}$$

$$\times \int_{\partial_j M \cap A} \int_{S(T_t \partial_j M^\perp)} \mathrm{Tr}^{T_t \partial_j M} \left(\widetilde{R}^m \widetilde{S}_{\nu_{N-j}}^{j-i-2m} \right)$$

$$\times \mathbb{1}_{\widehat{N_t M}}(-\nu_{N-j}) \, \mathcal{H}_{N-j-1}(d\nu_{N-j}) \mathcal{H}_j(dt), \tag{3.4.11}$$

and we define $\mathcal{L}_i(M;) \equiv 0$ if $i > \dim(M)$.

There is a lot to explain here. The constants are given by

$$C(m, i) \triangleq \begin{cases} \frac{(2\pi)^{i/2}}{s_{m+i}} & m+i > 0, \\ 1 & m = 0. \end{cases} \tag{3.4.12}$$

Integrals over empty regions and integrals associated with measures of negative index are taken to be zero. For the manifolds that we have been discussing, with no boundary or a smooth boundary $\partial_{N-1} M$, there are never more than two terms in the sum over j, but (3.4.11) is more general than this case. The terms $\widetilde{R}$ and $\widetilde{S}$ denote curvature tensors and second fundamental forms, respectively, and $\mathrm{Tr}^{T_t \partial_j M}$ denotes trace operators on the tangent spaces of $\partial_j M$.

Formula (3.4.11) simplifies somewhat if M is directly embedded in $\mathbb{R}^l$ and endowed with the canonical Riemannian structure on $\mathbb{R}^l$; i.e. $\widetilde{M} = \widehat{M} = \mathbb{R}^l$. Then, by the flatness of $\mathbb{R}^l$ ($R \equiv 0$ and so only the terms with $m = 0$ remain in (3.4.11)) the curvature measures (3.4.11) can be written as

$$\mathcal{L}_i(M, A) = \sum_{j=i}^{N} (2\pi)^{-(j-i)/2} C(N-j, j-i) \tag{3.4.13}$$

$$\times \int_{\partial_j M \cap A} \int_{S(\mathbb{R}^{N-j})} \frac{1}{(j-i)!} \mathrm{Tr}^{T_t \partial_j M} (\widetilde{S}_\eta^{j-i})$$

$$\times \mathbb{1}_{\widehat{N_t M}}(-\eta) \mathcal{H}_{N-j-1}(d\eta) \mathcal{H}_j(dt),$$

They simplify even further if M is a C^2 domain of $\mathbb{R}^N$. In this case $\mathcal{L}_N(M, U)$ is the Lebesgue measure of U and

$$\mathcal{L}_j(M, U) = \frac{1}{s_{N-j}(N-1-j)!} \int_{\partial M \cap U} \mathrm{Tr}(S_{\nu_t}^{N-1-j}) \, \mathrm{Vol}_{\partial M, g}, \tag{3.4.14}$$

for $0 \leq j \leq N-1$, where ν_t is the inward normal at t and $\mathrm{Vol}_{\partial M, g}$ is surface measure on ∂M.

In this setting (3.4.14) can also be written in another form that is often more conducive to computation. If we choose an orthonormal frame field $(E_1 \ldots, E_{N-1})$ on ∂M, and then extend this to one on M in such a way that E_N is the inward normal, then (3.4.14) can be rewritten as

$$\mathcal{L}_j(M, U) = \frac{1}{s_{N-j}} \int_{\partial M \cap U} \det r_{N-1-j}(\mathrm{Curv}) \, \mathrm{Vol}_{\partial M, g}, \qquad (3.4.15)$$

where, for a matrix A,

$$\det r_j(A) \overset{\Delta}{=} \text{Sum over all } j \times j \text{ principle minors of } A. \qquad (3.4.16)$$

and the curvature matrix Curv is given by

$$\mathrm{Curv}(i, j) \overset{\Delta}{=} S_{E_N}(E_i, E_j). \qquad (3.4.17)$$

It is important to note that while the elements of the curvature matrix may depend on the choice of basis, $\det r_{N-1-j}(\mathrm{Curv})$ is independent of the choice, as will be $\mathcal{L}_j(M, U)$.

Finally we define the (signed) total masses of the curvature measures to be the Lipschitz-Killing curvatures

$$\mathcal{L}_i(M) \overset{\Delta}{=} \mathcal{L}_i(M, M). \qquad (3.4.18)$$

The Lipschitz-Killing curvatures also appear under a variety of other names, such as quermassintegrales, Minkowski, Dehn and Steiner functionals, curvature invariants, and intrinsic volumes, although in many of these cases the ordering and normalisations are somewhat different from ours.

Among these the term *intrinsic volumes* and *curvature invariants* will have a special meaning for us and we shall return to these in a moment. In the meantime, however, here is the first tube formula:

Theorem 3.4.1 (Weyl's tube formula on $\mathbb{R}^l$). *Suppose $M \subset \mathbb{R}^l$ is regular stratified manifold. Then, for $\rho < \rho_c(M, \mathbb{R}^l)$,*

$$\mathcal{H}_l\left(\mathrm{Tube}(M, \rho)\right) = \sum_{i=0}^{\dim M} \rho^{l-i} \omega_{l-i} \mathcal{L}_i(M), \qquad (3.4.19)$$

where the $\mathcal{L}_j(M)$ are given by (3.4.13) and (3.4.18)

In preparation for Weyl's tube formula on spheres, we need to extend the Lipschitz-Killing measures slightly to obtain the one parameter family of measures

$$\mathcal{L}_i^{\kappa}(M, A) \triangleq \sum_{j=i}^{N} (2\pi)^{-(j-i)/2} \sum_{m=0}^{\lfloor \frac{j-i}{2} \rfloor} \frac{(-1)^m C(N-j, j-i-2m)}{m!(j-i-2m)!}$$

$$\times \int_{\partial_j M \cap A} \int_{S(T_t \partial_j M^{\perp})} \mathrm{Tr}^{T_t \partial_j M} \left(\left(\widetilde{R} + \frac{\kappa}{2} I^2 \right)^m \widetilde{S}_{\nu_{N-j}}^{j-i-2m} \right)$$

$$\times \mathbb{1}_{N_t M}(-\nu_{N-j}) \, \mathcal{H}_{N-j-1}(d\nu_{N-j}) \, \mathcal{H}_j(dt). \tag{3.4.20}$$

Note that $\mathcal{L}_i^0(M, \cdot) \equiv \mathcal{L}_i(M, \cdot)$.

As for $\mathcal{L}_i(M)$, we define the one parameter family of Lipschitz-Killing curvatures, or intrinsic volumes,

$$\mathcal{L}_i^{\kappa}(M) \triangleq \mathcal{L}_i^{\kappa}(M, M). \tag{3.4.21}$$

Furthermore, if $\kappa > 0$ and $\widetilde{M} = \widehat{M} = S_{\kappa^{-1/2}}(\mathbb{R}^N)$ $(\ell = N)$ then, as in (3.4.13), there is a simplification. This time it is due to the fact that then M will have constant negative curvature $-\kappa$, so that all the terms with $m \neq 0$ in (3.4.20) disappear. The result is

$$\mathcal{L}_i^{\kappa}(M, A) = \sum_{j=i}^{N} (2\pi)^{-(j-i)/2} C(N-1-j, j-i)$$

$$\times \int_{\partial_j M \cap A} \int_{S(T_t \partial_j M^{\perp})} \frac{1}{(j-i)!} \mathrm{Tr}^{T_t \partial_j M}(\widetilde{S}_{\eta}^{j-i})$$

$$\times \mathbb{1}_{\widehat{N_t M}}(-\eta) \mathcal{H}_{N-j-2}(d\eta). \tag{3.4.22}$$

It turns out that Euclidean and spherical Lipschitz-Killing curvatures are related, and expanding the trace term in (3.4.20) gives

$$\mathcal{L}_i^{\kappa}(\cdot) = \sum_{n=0}^{\infty} \frac{(-\kappa)^n}{(4\pi)^n} \frac{(i+2n)!}{n!i!} \mathcal{L}_{i+2n}(\cdot), \tag{3.4.23}$$

and

$$\mathcal{L}_i(\cdot) = \sum_{n=0}^{\infty} \frac{\kappa^n}{(4\pi)^n} \frac{(i+2n)!}{n!i!} \mathcal{L}_{i+2n}^{\kappa}(\cdot). \tag{3.4.24}$$

Here now is Weyl's tube formula on $S_\lambda(\mathbb{R}^l)$.

Theorem 3.4.2 (Weyl's tube formula on $S_\lambda(\mathbb{R}^l)$). *Suppose $M \subset S_\lambda(\mathbb{R}^l)$ is a regular stratified manifold. Then, for $\rho < \rho_c(M, S_\lambda(\mathbb{R}^l))$,*

$$\mathcal{H}_{l-1}\left(\mathrm{Tube}(M,\rho)\right)$$

$$= \sum_{i=0}^{\infty} \lambda^{l-1-i} G_{i,l-1-i}(\rho/\lambda)\mathcal{L}_i^{\lambda^{-2}}(M)$$

$$= \sum_{j=0}^{\infty} \left(\sum_{n=0}^{\lfloor \frac{j}{2} \rfloor} (-4\pi)^{-n} \frac{\lambda^{l-1+j}}{n!} \frac{j!}{(j-2n)!} G_{j-2n,l-1+2n-j}(\rho/\lambda) \right) \mathcal{L}_j(M),$$

where

$$G_{a,b}(\rho) \triangleq \frac{\pi^{b/2}}{b\Gamma\left(\frac{b}{2}+1\right)} \int_0^\rho \cos^a(r)\sin^{b-1}(r)\, dr$$

$$= \frac{\pi^{b/2}}{\Gamma\left(\frac{b}{2}\right)} \overline{IB}_{(a+1)/2,b/2}(\cos^2 \rho),$$

with

$$\overline{IB}_{(a+1)/2,b/2}(x) \triangleq \int_x^1 x^{(a-1)/2}(1-x)^{(b-2)/2}\, dx$$

the tail of the incomplete beta function.

3.4.3 Intrinsic Volumes

Here is a small section, which contains no more than two side comments which could have been included earlier. But the comments are important, and we want to make sure that you don't miss them, so they get their own bold faced, large font, header.

The first comment is that the Lipschitz-Killing curvatures, as we defined them in (3.4.11) for stratified manifolds embedded in $\mathbb{R}^N$ with the standard Euclidean metric, could have been defined, in precisely the same way, for any stratified Riemannian manifold. The curvature, second fundamental form and Hausdorff measure would then be the ones generated by the Riemannian metric, but the definition still makes sense. Thus, we shall do this, and define the Lipschitz-Killing curvatures of a general stratified Riemannian manifold by (3.4.11).

Of course, we can no longer claim that the same tube formula holds. Even in the cases of $\mathbb{R}^N$ and $S_\lambda(\mathbb{R}^l)$ we saw that while Lipschitz-Killing curvatures appeared in both, the constants involved were very much dependent on the physical differences between the two ambient spaces.

Having defined Lipschitz-Killing curvatures in general, there is an important point to note, and that is that the $\mathcal{L}_j$ are *intrinsic*, in the sense that they depend only on how M sits in $\widetilde{M}$ (via $\widetilde{R}$ and $\widetilde{S}$) but in no way reflect the largest manifold $\widehat{M}$. That this is the case is, at first sight, somewhat surprising, since they arose while deriving a tube formula for the tube in $\widehat{M}$. In fact, this was the deep part of Weyl's first proof of a general tube formula, and was why he considered what we have presented here – which does not prove that the $\mathcal{L}_j(M)$ are intrinsic – could have been "accomplished by any student in a course of calculus".

One consequence of this fact is that Lipschitz-Killing curvatures remain constant under *Riemannian isometries*. That is, suppose we have a diffeomorphism φ between two manifolds M and N. Suppose N has an associated Riemannian metric g, and we pull it back to M to generate a metric $\varphi_* g$ there. Then $(M, \varphi_* g)$ and (N, g) are indistinguishable from the point of view of Riemannian geometry, and, in particular, their Lipschitz-Killing curvatures, being intrinsic, are identical. This is one of the reasons that they are often called 'intrinsic volumes' or 'curvature invariants'.

A major consequence of this is related to what we hinted at earlier when talking about the mapping that took Gaussian processes with finite orthonormal expansions and on general parameter spaces to the canonical process on a subset of a sphere (cf. Sect. 2.4). For example, in Sect. 2.4.1 we saw that exceedence probabilities for such processes could be expressed in terms of exceedence probabilities of the canonical process. Assuming, correctly, that we can show that the latter depend only on the Lipschitz-Killing curvatures of subsets of the sphere, we now have a way of writing these probabilities in terms of the parameters of original random field.

Finally, we reiterate the perhaps surprising, and definitely deeper, fact that $\mathcal{L}_0(M)$ is the Euler characteristic of M, and so independent of any Riemannian structure. This is the celebrated Gauss–Bonnet Theorem.

Somewhat easier to check is the fact that $\mathcal{L}_N(M) = \mathcal{H}_N(M)$.

3.5 Probabilities of Tubes: Gaussian Minkowski Functionals

Our entire discussion of tube sizes has, up until now, been based on their volumes with respect to some Hausdorff measure, either Euclidean or determined by some Riemannian metric. However, there is an interesting extension of these ideas, from volumes to probabilities. This extension will be of crucial importance for us in what follows.

Suppose that $\mathbb{P}$ is a probability measure on $\mathbb{R}^k$ with an analytic density φ with respect to Lebesgue measure, a canonical example being the Gaussian measure γ_k corresponding to the distribution of a $N(0, I_{k\times k})$ random variable.

Then we can talk about the probability content of a tube, and, for A a regular, N-dimensional, stratified manifold in $\mathbb{R}^l$, replace (3.4.8) by

$$\mathbb{P}\{F_j(D_j(\rho))\} = \int_0^\rho \int_{S_j(r)} \mathbf{1}_{F_j(D_j(\rho))}\varphi(x)\,\mathcal{H}_{j,r}(dx)\,dr \qquad (3.5.1)$$

Now suppose we expand the density φ in a power series in normal directions and integrate over the hypersurfaces $S_j(r)$ of (3.4.9). Doing so gives an expansion[4] of the form

$$\mathbb{P}\{\text{Tube}(A,\rho)\} \stackrel{\Delta}{=} \sum_{j=0}^\infty \frac{\rho^j}{j!}\mathcal{M}_j^{\mathbb{P}}(A), \qquad (3.5.2)$$

in which the $\mathcal{M}_j^{\mathbb{P}}(A)$ can be represented as curvature integrals which, for the curious, we shall define in a moment. In general, however, finding explicit expressions for the $\mathcal{M}_j^{\mathbb{P}}$ via this construction is, not easy, and beyond the scope of these notes. Details, for Gaussian $\mathbb{P}$, can be found in Chap. 10 of RFG. When $\mathbb{P} = \gamma_k$, the distribution of a $N(0, I_{k \times k})$ random vector, then the corresponding $\mathcal{M}_j^{\gamma_k}$ are known as the *Gaussian Minkowski functionals*.

Alternatively, given any other way for computing the probability content of a tube, so that the left hand side of (3.5.2) is known, the right hand side gives an implicit definition of the $\mathcal{M}_j^{\mathbb{P}}$, much as we originally defined intrinsic volumes via Steiner's formula.

To see how this might work, consider the simple, one-dimensional example for which $\mathbb{P}$ is taken to be $\gamma = \gamma_1$, the distribution of a standard normal variable. For A we take the semi-infinite interval $[u, \infty)$. The argument is then simple and starts with a Taylor series expansion of the Gaussian distribution function Φ using Hermite polynomials and their property (2.2.7):

$$\gamma\big(\text{Tube}([u,\infty),\rho)\big) = 1 - \Phi(u - \rho)$$
$$= 1 - \left(\Phi(u) + \sum_{j=1}^\infty \frac{(-\rho)^j}{j!}\frac{(-1)^{j-1}}{\sqrt{2\pi}}H_{j-1}(u)e^{-u^2/2}\right)$$
$$= 1 - \Phi(u) + \sum_{j=1}^\infty \frac{\rho^j}{j!}\frac{1}{\sqrt{2\pi}}H_{j-1}(u)e^{-u^2/2},$$

so that, on comparison with (3.5.2), we find that

$$\mathcal{M}_j^\gamma([u,\infty)) = \frac{1}{\sqrt{2\pi}}H_{j-1}(u)e^{-u^2/2}, \qquad (3.5.3)$$

and we are done.

---

[4] Note that there is a qualitative difference between (3.5.2) and the Steiner and Weyl formula. The latter are expansions with only a finite number of terms, whereas (3.5.2) is, in principle, and generally in practice, an infinite expansion.

Another difference lies in the fact that $\mathcal{M}_j^{\mathbb{P}}(A) \equiv 0$ for all $j < N - \dim(A)$, so that the sum in (3.5.2) actually starts at $j = N - \dim(A)$.

This was a particularly easy computation, but it turns out that in very many of the important cases to follow, in which $\mathbb{P} = \gamma_k$, similar arguments hold. For example, suppose that $F : \mathbb{R}^k \to \mathbb{R}$ is smooth enough for the sets $F^{-1}([u, \infty))$ to be smooth and locally convex. Suppose also that

$$\text{Tube}\left(F^{-1}([u, \infty)), \rho\right) = F^{-1}([u - \rho, \infty)), \qquad (3.5.4)$$

a relationship that we shall see holds surprisingly often in practice. Then it follows immediately from the tube formula (3.5.2) that, for A of this form,

$$\mathcal{M}_k^{\mathbb{P}}(A) = \frac{d^k}{d\rho^k}\mathbb{P}\left\{F^{-1}([u - \rho, \infty))\right\}\Big|_{\rho=0}$$

$$= (-1)^k \frac{d^k}{dx^k}\mathbb{P}\left\{F(Z) \geq x\right\}\Big|_{x=u}, \qquad (3.5.5)$$

where $Z \sim \gamma_k$.

These kinds of sets appear often in applications, in which they take the form of rejection regions of a statistical test. In particular, given this result, it should now be clearer why we initially introduced the structure of Fig. 1.1.1, in which considered random fields of the form $g = F \circ f$ for vector valued, Gaussian f. The above observations are going to allow us to compute mean properties of the excursion sets of the non-Gaussian g from the Gaussian kinematic formula. Some examples are given in Sect. 5.2.

Before moving on, we devote a half page to the curious reader, who wants to see an explicit construction of the $\mathcal{M}_j^{\gamma_k}$, rather than relying on their implicit definition via the expansion (3.5.2). Firstly, however, we need to somewhat extend the notion of Lipschitz-Killing curvatures.

Take Borel $A \subset \mathbb{R}^l$ and $B \subset S(\mathbb{R}^l)$, retain the notation of (3.4.11) and define, for $0 \leq i \leq l - 1$, a family of *generalised Lipschitz-Killing curvature measures* supported on $M \times S(\mathbb{R}^l)$ by

$$\widetilde{\mathcal{L}}_i(M, A \times B) \triangleq \sum_{j=i}^{l}(2\pi)^{-(j-i)/2}\sum_{m=0}^{\lfloor\frac{j-i}{2}\rfloor}\frac{(-1)^m C(l-j, j-i-2m)}{m!\,(j-i-2m)!}$$

$$\times \int_{\partial_j M \cap A}\int_{S(T^\perp \partial_j M) \cap B}\text{Tr}^{T_t \partial_j M}\left(\widetilde{R}^m \widetilde{S}_{\nu_{l-j}}^{j-i-2m}\right)$$

$$\times \mathcal{H}_{l-j-1}(d\nu_{l-j})\mathcal{H}_j(dt). \qquad (3.5.6)$$

For $i = l$ we define $\widetilde{\mathcal{L}}_i$ only on sets of the form $A \times S(\mathbb{R}^l)$ by setting $\widetilde{\mathcal{L}}_i(A \times S(\mathbb{R}^l)) = \mathcal{H}_l(A)$. For Borel $f : \mathbb{R}^l \times S(\mathbb{R}^l) \to \mathbb{R}$, let $\widetilde{\mathcal{L}}_i(M, f)$ denote the integral of f with respect to $\widetilde{\mathcal{L}}_i$.

A change of numbering and normalization now defines the *generalised Minkowski curvature measures* as

$$\widetilde{\mathcal{M}}_j(M, A \times B) \triangleq (j! \, \omega_j) \, \widetilde{\mathcal{L}}_{l-j}(M, A \times B). \tag{3.5.7}$$

With these definitions we can now give a direct definition of the Gaussian Minkowski functionals appearing of (3.5.2) by setting

$$\mathcal{M}_j^\gamma(M) \triangleq (2\pi)^{-l/2} \sum_{m=0}^{j-1} \binom{j-1}{m} \widetilde{\mathcal{M}}_{m+1}\left(M, H_{j-1-m}\big(\langle t, \eta \rangle\big) \, e^{-|t|^2/2}\right). \tag{3.5.8}$$

For more details, see *RFG*.

3.6 Kinematic Formulae

We already met the kinematic fundamental formula for nice subsets of $\mathbb{R}^N$ back in Sect. 1.3.3, cf. (1.3.9). All that remains to say about it is that back there we were not very precise about the classes of sets for which it held. Now we have the language to tell you that it holds, among others, for regular stratified manifolds in $\mathbb{R}^N$, a proof of which can be found, for example, in [16].

In fact, it holds in far greater generality, and for a full treatment of this important result in a variety of scenarios you should turn to any of the classic references, including [13, 22, 38, 57, 71, 73].

What will be far more important for us, however, is a version of the kinematic fundamental formula for subsets of $S_{\sqrt{n}}(\mathbb{R}^n)$, in which the averaging is carried out over $G_{n,\lambda}$, the group of isometries (i.e. rotations) on $S_\lambda(\mathbb{R}^n)$.

Noting that $G_{n,\lambda} \simeq O(n)$, we normalize Haar measure $\nu_{n,\lambda}$ on $G_{n,\lambda}$ so that, for any $x \in S_\lambda(\mathbb{R}^n)$ and every Borel $A \subset S_\lambda(\mathbb{R}^n)$, we have

$$\nu_{n,\lambda}\left(\{g_n \in G_{n,\lambda} : g_n x \in A\}\right) = \mathcal{H}_{n-1}(A). \tag{3.6.1}$$

The kinematic fundamental formula on $S_\lambda(\mathbb{R}^n)$ then reads as follows, where M_1 and M_2 are regular stratified manifolds in $S_\lambda(\mathbb{R}^n)$.

$$\int_{G_{n,\lambda}} \mathcal{L}_i^\lambda\left(M_1 \cap g_n M_2\right) d\nu_{n,\lambda}(g_n)$$

$$= \sum_{j=0}^{n-1-i} \begin{bmatrix} i+j \\ i \end{bmatrix} \begin{bmatrix} n-1 \\ j \end{bmatrix}^{-1} \mathcal{L}_{i+j}^\lambda(M_1)\mathcal{L}_{n-1-j}^\lambda(M_2)$$

$$= \sum_{j=0}^{n-1-i} \frac{s_{i+1}s_n}{s_{i+j+1}s_{n-j}} \mathcal{L}_{i+j}^\lambda(M_1)\mathcal{L}_{n-1-j}^\lambda(M_2), \tag{3.6.2}$$

where the functionals $\mathcal{L}_i^\lambda(\cdot)$ are from the one parameter family defined in (3.4.20).

3.7 Crofton's Formula

A result closely related to the kinematic fundamental formula is a much older result due to Morgan Crofton and named for him, in which instead of looking at the intersections between two compact sets one looks at 'random' cross-sections, of various dimensions, of a single set. Crofton's formula will help us later on a couple fronts. The most important will be that it shows that one can say much about Lipschitz-Killing curvatures by knowing only about Euler characteristics of cross-sections.

To formulate Crofton's formula we need to recall the *affine Grassmanian* manifold $\mathrm{Graff}(N,k)$, the set of k-dimensional subspaces of $\mathbb{R}^N$; *viz.* the collection of all linear spaces in $\mathbb{R}^N$ that do not necessarily pass through the origin. Noting that the affine Grassmanian is diffeomorphic to $\mathrm{Gr}(N,k) \times \mathbb{R}^N$, it has a natural measure, λ_k^N say, which factors as Haar measure ν_k^N on the Grassmanian $\mathrm{Gr}(N,k)$ of all linear subspaces of $\mathbb{R}^N$ (which must pass through the origin) and Lebesque measure on $\mathbb{R}^N$. We normalize ν_k^N so that

$$\nu_k^N(\mathrm{Gr}(N,k)) = \begin{bmatrix} N \\ k \end{bmatrix}.$$

Crofton's formula then states that, for a regular stratified manifold $M \in \mathbb{R}^N$,

$$\int_{\mathrm{Graff}(N,N-k)} \mathcal{L}_j(M \cap V)\, d\lambda_{N-k}^N(V) = \begin{bmatrix} k+j \\ j \end{bmatrix} \mathcal{L}_{k+j}(M). \qquad (3.7.1)$$

The $\mathcal{L}_j$ in (3.7.1) are all computed with respect to the standard Euclidean metric on $\mathbb{R}^N$.

The special case $k = 0$ of Crofton's formula is generally known as *Hadwiger's formula* and is given by

$$\mathcal{L}_k(M) = \int_{\mathrm{Graff}(N,N-k)} \mathcal{L}_0(M \cap V)\, d\lambda_{N-k}^N(V). \qquad (3.7.2)$$

Note how, as promised, Hadwiger's formula allows one to compute all the Lipschitz-Killing curvatures from the Euler characteristics of cross-sections.

Nice places to read about these results, in the setting of M in the convex ring, are [57, 71], while a full proof in the generality we require can be found in [16].

3.8 Morse's Theorem

Crofton's formula tells us that, among all the Lipschitz-Killing curvatures, there is something special about the first one $\mathcal{L}_0(M) \equiv \varphi(M)$, the Euler characteristic. We already noticed this back in the Introduction, when we started with the Euler characteristics and saw a number of way in which it could be defined and/or calculated, depending on which area of geometry or topology one approached it. One of the ways was via the alternating sum (1.2.4) of number of critical points of different types.

In this section we want to state that result more formally, and in the setting of stratified Riemannian manifolds.

Before we start, we need to define Morse functions, and remind you that our regular stratified manifold M must, by the definition of regularity, be embedded in a C^3 manifold $\widetilde{M}$. Then a function $f \in C^2(\widetilde{M})$ is called a *Morse function* on M if it satisfies the following two conditions on each stratum $\partial_j M$, $j = 0, \ldots, \dim(M)$.

(a) All the critical points of $f_{|\partial_j M}$ on $\partial_j M$ (i.e. the points $t \in \partial_j M$ for which the covariant derivative $\nabla \widetilde{f}_t \in T_t^{\perp} \partial_j M$) are *non-degenerate*, in the sense that at these points the covariant Hessian $\nabla^2 f_{|T_t \partial_j M}$ is non-degenerate when considered as a bilinear mapping.

(b) The restriction of f to $\overline{\partial_j M} = \bigcup_{i=0}^{j} \partial_i M$ has no critical points on $\bigcup_{i=0}^{j-1} \partial_i M$.

Now, for a Morse function f on M, define the *Morse index* $\iota_{f, \partial_j M}(t)$ of a critical point $t \in \partial_j M$ of $f_{|M}$ to be the dimension of the largest subspace L of $T_t \partial_j M$ such that $\nabla^2 f(t)\big|_L$ is negative definite. Thus, a point of tangential Morse index zero is a local minimum of f on $\partial_j M$, while a point of index j is a local maximum. Other indices correspond to saddle points of various kinds. Here then is Morse's theorem, due, in the form below, to Goresky and MacPherson [43]. A simple but important special case (especially for the reader who wants to assume that stratified manifolds are all cubes) will follow.

Theorem 3.8.1 (Morse's Theorem). *Let M be a regular stratified manifold embedded in $\widetilde{M}$ and $\widetilde{f} \in C^2(\widetilde{M})$ a Morse function on M. Then, setting $f = \widetilde{f}_{|M}$,*

$$\varphi(M) = \sum_{j=0}^{N} \sum_{\{t \in \partial_j M : \nabla f_t \in T_t^{\perp} \partial_j M\}} (-1)^{\iota_{f, \partial_j M}(t)} \mathbb{1}_{\{-\nabla f_t \in N_t M\}}, \qquad (3.8.1)$$

where $\varphi(M)$ is the Euler characteristic of M.

Note that if M is does not have a boundary, then this is precisely the formula we gave in the Introduction for computing Euler characteristics via critical points, *viz.* (1.2.4).

There is an extension of Morse's formula to cover the Euler characteristics of excursion sets, a proof of which (assuming Morse's theorem) can be found in *RFG*.

Corollary 3.8.2. *Let f and M be as in Theorem 3.8.1, and suppose that $u \in \mathbb{R}$ is not a critical value of $f_{|\partial_j M}$ for any $j = 0, \ldots, N$. Then, writing $f = \widetilde{f}_{|M}$,*

$$\varphi\left(M \cap f^{-1}[u, \infty)\right) = \sum_{j=0}^{N} \sum_{\{t \in \partial_j M : f_t > u, \nabla f_t \in T_t^{\perp} \partial_j M\}} (-1)^{\iota - f, \partial_j M(t)} \mathbf{1}_{\{\nabla f_t \in N_t M\}}.$$
(3.8.2)

We call the points counted in the above sum – i.e. those for which $\nabla f_t \in N_t(M)$ – the *extended outward critical points* of f.

Morse's Theorem is a deep and important result, requiring concepts from both differential and algebraic topology for its proof. It is actually somewhat more general than as stated above, since in its full form it also gives a series of inequalities linking Betti numbers of different orders to numbers of critical points of different index. We shall return to this later, but now look at the special case alluded to above.

If M is simply the N-dimensional cube $I^N = [0, 1]^N$, then the Morse representation (3.8.2) of the excursion set Euler characteristic becomes very simple.

Let $\mathcal{J}_k \equiv \partial_k I^N$ denote the collection of faces of dimension k in I^N. Then we can rewrite the sum (3.8.2) as

$$\varphi\left(A_u(f, I^N)\right) = \sum_{k=0}^{N} \sum_{J \in \mathcal{J}_k} \sum_{i=0}^{k} (-1)^i \mu_i(J),$$
(3.8.3)

where, for $i \leq \dim(J)$,

$$\mu_i(J) \stackrel{\Delta}{=} \# \left\{t \in J : f(t) > u, \nabla f_{|J}(t) = 0, \nabla f_t \in N_t I^N, \iota_{-f, J}(t) = i\right\}.$$

Note that to each face $J \in \mathcal{J}_k$ there corresponds a subset $\sigma(J)$ of $\{1, \ldots, N\}$, of size k, and a sequence of $N - k$ zeroes and ones $\varepsilon(J) = \{\varepsilon_1, \ldots, \varepsilon_{N-k}\}$ so that

$$J = \left\{t \in I^N : t_j = \varepsilon_j, \text{ if } j \notin \sigma(J), \ 0 < t_j < 1, \text{ if } j \in \sigma(J)\right\}.$$

Setting $\varepsilon_j^* = 2\varepsilon_j - 1$, it is now not hard to see that $\mu_i(J)$ is given by the number of points $t \in J$ satisfying the following four conditions:

$$f(t) \geq u,$$
(3.8.4)
$$f_j(t) = 0, \qquad j \in \sigma(J)$$
(3.8.5)

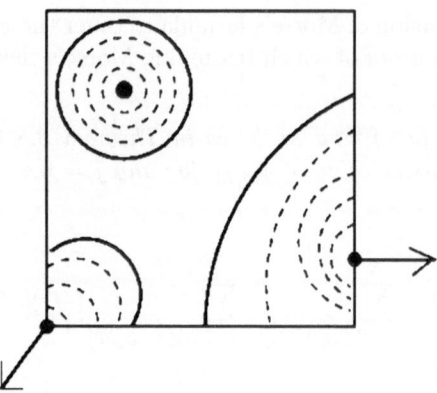

Fig. 3.8.1 Points contributing to the Euler characteristic of an excursion set, along with their associated outward normals

$$\varepsilon_j^* f_j(t) > 0, \qquad j \notin \sigma(J) \qquad\qquad (3.8.6)$$

$$\text{index}\,(f_{mn}(t))_{(m,n\in\sigma(J))} = k - i, \qquad\qquad (3.8.7)$$

where, as usual, subscripts denote partial differentiation, and, consistent with the definition of the index of a critical point, we define the *index of a matrix* to be the number of its negative eigenvalues.

In Fig. 3.8.1 there are three points which contribute to the Euler characteristic of $A_u(f, I^2)$, which itself is made up of three disjoint pieces. One, in the centre of the upper left disk, contributes via $J = (I^2)^\circ = \mathcal{J}_2$. That on the right side contributes via $J =$ 'right side'$\in \mathcal{J}_1$, and that on the lower left corner via $J = \{0\} \in \mathcal{J}_0$.

Chapter 4
The Gaussian Kinematic Formula

Our aim in this chapter is to show you how to prove the Gaussian kinematic formula, (1.3.3). That is, that for $f : M \to \mathbb{R}^d$, with component random fields $f_1, \ldots, f_d$ which are smooth, zero mean, unit variance, and Gaussian, and where M and $D \subset \mathbb{R}^d$ are nice enough,

$$\mathbb{E}\left\{\mathcal{L}_i\left(A(f, M, D)\right)\right\} = \sum_{j=0}^{\dim M - i} \begin{bmatrix} i+j \\ j \end{bmatrix} (2\pi)^{-j/2} \mathcal{L}_{i+j}(M) \, \mathcal{M}_j^\gamma(D). \quad (4.0.1)$$

The $\mathcal{L}_j$ are the Lipschitz-Killing curvatures, or intrinsic volumes, of Sect. 3.4.3, computed with respect to a Riemannian metric induced on M by the component random fields f_j. These will soon be defined in Sect. 4.5.

Of course "to show you how to prove" is not quite the same as "to prove", and, indeed, we shall not attempt the latter. For this you will have to read Chap. 15 of *RFG*. What we shall do, however, is to prove a special case of (4.0.1), showing that for real valued, mean zero, unit variance, smooth, isotropic, Gaussian random fields on rectangles $T = \prod_1^N [0, T_j]$,

$$\mathbb{E}\left\{\mathcal{L}_0\left(A(f, M, [u, \infty))\right)\right\} = \mathbb{E}\left\{\varphi\left(A_u\right)\right\}$$
$$= e^{-u^2/2} \sum_{j=0}^{N} (2\pi)^{-(j+1)/2} \lambda_2^{j/2} \mathcal{L}_k(T) H_{j-1}(u).$$
$$(4.0.2)$$

The Hermite polynomials H_j were defined at (2.2.5)–(2.2.6), and λ_2 is the second spectral moment (2.6.9) of f. The $\mathcal{L}_j$ in (4.0.2) are the standard, Euclidean, Lipschitz-Killing curvatures, which in this case are given by

$$\mathcal{L}_j(T) = \sum_{j_1 \ldots j_k} T_{j_1} \cdots T_{j_k}, \quad (4.0.3)$$

the sum taken over the $\binom{N}{k}$ *distinct* choices of k indices between 1 and N. (cf. (1.2.12).) $\mathcal{L}_0$ is, as always, the Euler characteristic φ.

R.J. Adler and J.E. Taylor, *Topological Complexity of Smooth Random Functions*, Lecture Notes in Mathematics 2019, DOI 10.1007/978-3-642-19580-8_4, © Springer-Verlag Berlin Heidelberg 2011

The proof of (4.0.2) is not hard (particularly as we shall also skip some of the messy but conceptually straightforward calculus that it involves). The proof is also key to the first part of the proof of the general case, and will be given in Sect. 4.2. It will take us the remainder of the chapter to get to the most general case, and we shall proceed in a number of small steps, each of independent interest, rather than trying to make one giant leap.

The first steps involve moving from the Euler characteristic to the other Lipschitz-Killing curvatures in Sect. 4.3, and then from isotropy to stationarity in Sect. 4.4. Life starts getting complicated after this, and so in Sect. 4.5 we explain the need for a Riemannian approach to the problem based on the 'induced metric'.

Section 4.6 treats another special case, that of the canonical isotropic process on the N-sphere, which is extended to quite general random fields, but with finite expansions, in Sect. 4.7. The final Sect. 4.8 then describes how to put all the pieces together so as to prove the most general results, although to fill in the details you will need to go to *RFG*.

However, before we start on this path, we need an important technical result of significant independent interest.

4.1 The Kac-Rice Metatheorem

The technical result we need has its roots going back at least 70 years, to seminal papers by Rice [69, 70] and Kac [54] who were interested in the number of real zeroes and the number of maxima of finite order polynomials with Gaussian coefficients. The more modern versions still retain the names of the pioneers, although we shall often call the version we need the Kac-Rice (expectation) metatheorem rather than the Kac-Rice formula.

To state it, consider two vector valued random fields $f = (f^1, \ldots, f^N)$ and $g = (g^1, \ldots, g^K)$ defined on some compact set $M \subset \mathbb{R}^N$ with non-empty interior. For $B \subset \mathbb{R}^K$, we want to compute the expectations

$$\mathbb{E}\{\#\{t \in M : f(t) = u, \ g(t) \in B\}\}. \tag{4.1.1}$$

Perhaps the most basic application of (4.1.1) is to prove the famous Rice formula for the expected number of upcrossings of the level u of a real valued process f on the line, where an upcrossing is defined as a point t where $f(t) = u$ and $f(t)$ is increasing. In this example $g = \dot{f}$ and $B = [0, +\infty)$.

We, however, shall generally use it to compute expected numbers of critical points of real valued fields h, so that $f = \nabla h$ and, if the critical points to be counted are, for example, local maxima, then g would be the indicator of the event that $\nabla^2 h$ is a negative definite matrix and $B = \{1\}$.

The theorem follows. Note that since ∇f denotes the gradient of f, and f takes values in $\mathbb{R}^N$, the gradient is the $N \times N$ matrix of first-order partial derivatives of f; i.e.

$$(\nabla f)(t) \equiv \nabla f(t) \equiv \left(f_j^i(t)\right)_{i,j=1,\ldots,N} \equiv \left(\frac{\partial f^i(t)}{\partial t_j}\right)_{i,j=1,\ldots,N}.$$

Theorem 4.1.1. *Let f, g, M and B be as above, with the additional assumption that the boundaries of M and B have finite $N-1$ and $K-1$ dimensional measures, respectively. Furthermore, assume that the following conditions are satisfied for some $u \in \mathbb{R}^N$:*

(a) All components of f, ∇f, and g are a.s. continuous and have finite variances (over M).

(b) For all $t \in M$, the marginal densities $p_t(x)$ of $f(t)$ (implicitly assumed to exist) are continuous at $x = u$.

(c) The conditional densities $p_t(x|\nabla f(t), g(t))$ of $f(t)$ given $g(t)$ and $\nabla f(t)$ (implicitly assumed to exist) are bounded above and continuous at $x = u$, uniformly in $t \in M$.

(d) The conditional densities $p_t(z|f(t) = x)$ of $\det \nabla f(t)$ given $f(t) = x$, are continuous for z and x in neighbourhoods of 0 and u, respectively, uniformly in $t \in M$.

(e) The conditional densities $p_t(z|f(t) = x)$ of $g(t)$ given $f(t) = x$, are continuous for all z and for x in a neighbourhood u, uniformly in $t \in M$.

(f) The following moment condition holds:

$$\sup_{t \in M} \max_{1 \leq i,j \leq N} \mathbb{E}\left\{\left|f_j^i(t)\right|^N\right\} < \infty. \tag{4.1.2}$$

(g) The moduli of continuity with respect to the usual Euclidean norm (cf. (2.5.9)) of each of the components of f, ∇f, and g satisfy

$$\mathbb{P}\{\omega(\eta) > \varepsilon\} = o\left(\eta^N\right), \qquad \text{as } \eta \downarrow 0, \tag{4.1.3}$$

for any $\varepsilon > 0$.

Then, if

$$N_u \equiv N_u(M) \equiv N_u(f, g : M, B)$$

denotes the number of points in M for which

$$f(t) = u \in \mathbb{R}^N \quad \text{and} \quad g(t) \in B \subset \mathbb{R}^K,$$

and $p_t(x, \nabla y, v)$ denotes the joint density of $(f_t, \nabla f_t, g_t)$, we have, with $D = N^2 + K$,

$$\mathbb{E}\{N_u\} = \int_M \int_{\mathbb{R}^D} |\det \nabla y| \, \mathbf{1}_B(v) \, p_t(u, \nabla y, v) \, d(\nabla y) \, dv \, dt. \tag{4.1.4}$$

It is sometimes more convenient to write this as

$$\mathbb{E}\{N_u\} = \int_M \mathbb{E}\left\{ |\det \nabla f(t)| \, \mathbb{1}_B(g(t)) \,\Big|\, f(t) = u \right\} p_t(u)\, dt, \quad (4.1.5)$$

where p_t here is the density of $f(t)$.

In our applications of Theorem 4.1.1, the expression (4.1.5) will be the principal form used.

For a full proof, see *RFG*, where this result appears as Theorem 11.2.1. An outline of the beginning of this proof, that at least shows from where the result comes, will be given in a moment.

Conditions (a)–(g) are often tedious to check, but almost disappear when both f and g are either Gaussian, or simple functions of vector valued Gaussian random fields. In these situations, the primary consideration becomes one of sample path continuity and differentiability, which we already looked at in Chap. 2.

Corollary 4.1.2. *Let f and g be centered Gaussian fields with M and B satisfying the conditions of Theorem 4.1.1. Assume that, for each $t \in M$, the joint distribution of $(f(t), \nabla f(t), g(t))$ is non-degenerate.*

Write $C_f^i = C_f^i(s,t)$ for the covariance function of f^i, $C_{f_j}^i = \partial^2 C_f^i / \partial s_j \partial t_j$ for the covariance function of $f_j^i = \partial f^i / \partial t_j$, and C_g^i for the covariance function of g^i. If

$$\max_{i,j} \left| C_{f_j}^i(t,t) + C_{f_j}^i(s,s) - 2C_{f_j}^i(s,t) \right| \leq K \left| \ln |t - s| \right|^{-(1+\alpha)},$$
$$\max_i \left| C_g^i(t,t) + C_g^i(s,s) - 2C_g^i(s,t) \right| \leq K \left| \ln |t - s| \right|^{-(1+\alpha)}, \quad (4.1.6)$$

for some finite $K > 0$, some $\alpha > 0$ and all $|t - s|$ small enough, then the conclusions of Theorem 4.1.1 hold.

Outline of a proof of Theorem 4.1.1. We start with $\delta_\varepsilon : \mathbb{R}^N \to \mathbb{R}$ an approximate delta function, constant on the N-ball $B_\varepsilon(\mathbb{R}^N) = \{t \in \mathbb{R}^N : |t| < \varepsilon\}$, zero elsewhere, and normalized so that

$$\int_{B(\varepsilon)} \delta_\varepsilon(t)\, dt = 1. \quad (4.1.7)$$

We then claim that

$$N_u(f, g; M, B) = \lim_{\varepsilon \to 0} \int_M \delta_\varepsilon(f(t) - u)\, \mathbb{1}_B(g(t))\, |\det \nabla f(t)|\, dt. \quad (4.1.8)$$

If this is true, then, with no further pretense to rigor, take expectations on both sides and freely change the orders of limit and expectation to find that

$$\mathbb{E}\{N_u\} = \lim_{\varepsilon \to 0} \mathbb{E} \int_M \delta_\varepsilon(f(t) - u)\, \mathbf{1}_B(g(t))\, |\det \nabla f(t)|\, dt$$

$$= \int_M \int_{\mathbb{R}^{N^2}} \int_{\mathbb{R}^K} \mathbf{1}_B(v)\, |\det \nabla y|$$

$$\times \left\{ \lim_{\varepsilon \to 0} \int_{\mathbb{R}^N} \delta_\varepsilon(x - u)\, p_t(x, \nabla y, v)\, dx \right\} d\nabla y\, dv\, dt,$$

where the p_t are the obvious densities. Taking the limit in the innermost integral yields

$$\mathbb{E}\{N_u\} = \int_M \int \int \mathbf{1}_B(v)\, |\det \nabla y|\, p_t(u, \nabla y, v)\, d\nabla y\, dv\, dt$$

$$= \int_M \mathbb{E}\{|\det \nabla f(t)|\, \mathbf{1}_B(g(t))\,|\, f(t) = u\}\, p_t(u)\, dt,$$

which is what we wanted to show.

Of course, interchanging the order of integration and the limiting procedure requires justification, and not only is it far from trivial it is in fact so hard to do that a fully rigorous proof requires a rather different approach. Again, see the proof of Theorem 11.2.1 of *RFG* for details. Nevertheless, we shall assume that, under the conditions of the theorem, the interchange works.

Thus, all that remains is to prove (4.1.8).

To save on notation, and without any loss of generality, we take $u = 0$. Consider those $t \in M$ for which $f(t) = 0$, of which we claim (without proof) that under the conditions of the theorem there is, a.s., only a finite number and that none lie in ∂M. Consequently, each one can be surrounded by an open ball, of radius η, say, in such a way that the balls neither overlap nor intersect ∂M. Furthermore, we can take η small enough so that within each ball $g(t)$ always lies in either B or the interior of its complement, but never both.

Let $\sigma(\varepsilon)$ be the ball $|f| < \varepsilon$ in the image space $\mathbb{R}^k$ of f. From what we have just claimed follows the fact that we can also choose ε small enough for the inverse image of $\sigma(\varepsilon)$ in M to be contained within the union of the η spheres.

Furthermore, by the inverse mapping theorem, we can choose ε, η so small that, for each η sphere in M, $\sigma(\varepsilon)$ is contained in the f image of the η sphere. The restriction of f to such a sphere will be one-one. Since the Jacobian of the mapping of each η sphere by f is $|\det \nabla f(t)|$ it follows that we can choose ε small enough so that

$$N_0 = \int_M \delta_\varepsilon(f(t))\, \mathbf{1}_B(g(t))\, |\det \nabla f(t)|\, dt.$$

This follows since each η sphere in M over which $g(t) \in B$ will contribute exactly one unit to the integral, while all points outside the η spheres will not be mapped onto $\sigma(\varepsilon)$. Since the left-hand side of this expression is independent of ε we can take the limit as $\varepsilon \to 0$ to obtain (4.1.8), as required.   □

There is another way to prove Theorem 4.1.1 that has also been around, in one form or another, for many years, and is based on Federer's coarea formula, which we met in Sect. 3.2. The French among you may prefer it. Rewrite the coarea formula for the case (3.2.4) as

$$\int_{\mathbb{R}^N} \left(\sum_{t: f(t)=u} \alpha(t) \right) du = \int_{\mathbb{R}^N} \alpha(t) \, |\det \nabla f(t)| \, dt,$$

assuming that f and $\alpha : \mathbb{R}^N \to \mathbb{R}^N$ are sufficiently smooth.

Take $\alpha(t) = \varphi(f(t)) \mathbb{1}_M(t)$, where φ is a smooth (test) function. (Of course, α is now no longer smooth, but we shall ignore this for the moment.) The above then becomes

$$\int_{\mathbb{R}^N} \varphi(u) \, N_u(f : M) \, du = \int_M \varphi(f(t)) \, |\det \nabla f(t)| \, dt.$$

Now take expectations (assuming this is allowed) of both sides to obtain

$$\int_{\mathbb{R}^N} \varphi(u) \, \mathbb{E}\{N_u(f : M)\} \, du$$
$$= \int_M \mathbb{E}\{\varphi(f(t)) \, |\det \nabla f(t)|\} \, dt$$
$$= \int_{\mathbb{R}^N} \varphi(u) \int_M \mathbb{E}\{|\det \nabla f(t)| \, | \, f(t) = u\} \, p_t(u) \, dt du.$$

Since φ was arbitrary, this implies that for (Lebesgue) almost every u,

$$\mathbb{E}\{N_u(f : M)\} = \int_M \mathbb{E}\{|\det \nabla f(t)| \, | \, f(t) = u\} \, p_t(u) \, dt, \quad (4.1.9)$$

which is precisely (4.1.5) of Theorem 4.1.1 with the g there identically equal to 1. Modulo this restriction on g, which is simple to remove, this is what we need. The problem, however, is that since it is true only for almost every u one cannot be certain that it is true for a specific value of u.

To complete the proof, we need only show that both sides of (4.1.9) are continuous functions of u and that the assumptions of convenience made above are no more than that. This, of course, is not as trivial as it may sound, and going through the arguments actually leads to virtually the same long list of conditions that appeared in the statement of the theorem. See [11] for the details.

While not at all obvious at first sight, hidden away in Theorem 4.1.1 is another result, about higher moments of $N_u(f, g; B, M)$. In particular, under additional conditions, it can be shown that

$$\mathbb{E}\{N_u (N_u - 1) \cdots (N_u - k + 1)\}$$

$$= \int_{M^k} \mathbb{E}\Big\{ \prod_{j=1}^k |\det \nabla f(t_j)| \, \mathbb{1}_B(g(t_j)) \, \Big| \, \bar{f}(\bar{t}) = \bar{u} \Big\} p_{\bar{t}}(\bar{u}) \, d\bar{t},$$

where $\bar{t} = (t_1, \ldots, t_k)$, $\bar{f}(\bar{t}) = (f(t_1), \ldots, f(t_k))$, and $p_{\bar{t}}$ is the probability density of $\bar{f}(\bar{t})$. Since we shall not need this result in what follows, we send you to *RFG* or [11] for further details.

4.2 Real Isotropic Fields on Rectangles: Euler Characteristic

Our aim in this section is to give a proof of (4.0.2), which is the special case of the GKF for real isotropic Gaussian fields on rectangles, and looks only at the mean Euler characteristic of excursion sets. Modulo some calculus which is left to you, the proof is complete and self-contained, as is the statement, for the reader who might want only this result without having to chase back through the notes for notation.

Theorem 4.2.1. *Let f be a real valued, centered, isotropic Gaussian field on $\mathbb{R}^N$ with variance σ^2. Denote the first and second order partial derivatives of f by f_i and f_{ij}, with ∇f the vector of the f_i and $\nabla^2 f$ the matrix of the f_{ij}. Assume that*

(i) The joint distributions of $(f(t), \nabla f(t), \nabla^2 f(t))$ are, for each $t \in T$, non-degenerate.

(ii) If $C_{ij}(s, t)$ is the covariance function of f_{ij}, then

$$\max_{i,j} |C_{ij}(t, t) + C_{ij}(s, s) - 2C_{ij}(s, t)| \leq K |\ln |t - s||^{-(1+\alpha)},$$

for some finite $K > 0$, some $\alpha > 0$, and all $|t - s|$ small enough.

Let $T = \prod_1^N [0, T_j]$, and for finite, real u, consider the excursion set $A_u \equiv A_u(f, T)$. With φ denoting the Euler characteristic,

$$\mathbb{E}\{\varphi(A_u)\} = e^{-u^2/2\sigma^2} \sum_{k=0}^N \frac{\lambda_2^{k/2}}{(2\pi)^{(k+1)/2} \sigma^k} \mathcal{L}_k(T) H_{k-1}\left(\frac{u}{\sigma}\right), \quad (4.2.1)$$

where the Hermite polynomials H_j were defined at (2.2.5)–(2.2.6), and λ_2 is the second spectral moment (2.6.9) of f. The $\mathcal{L}_j$ are the standard, Euclidean, Lipschitz-Killing curvatures

$$\mathcal{L}_j(T) = \sum_{j_1 \cdots j_k} T_{j_1} \cdots T_{j_k}, \qquad (4.2.2)$$

the sum taken over the $\binom{N}{k}$ distinct choices of k indices between 1 and N. (cf. (1.2.12).)

Before starting the proof of Theorem 4.2.1 we recall from Sect. 3.8 that there is a way to write $\varphi(A_u(f, T))$ via the number of critical points of f of various types on the various faces of T. This leads us to some notation:

Firstly, we write $\partial_k T$ for the collection of the $2^{N-k}\binom{N}{k}$ faces of dimension k in T. As opposed to our previous conventions, in this chapter we take these faces as closed. Thus all faces in $\partial_k T$ are subsets of faces in $\partial_{k'} T$ for all $k' > k$.

Each k-dimensional face $J \in \partial_k T$ is determined by a subset $\sigma(J)$ of $\{1, \ldots, N\}$, of size k, and a sequence of $N - k$ zeros and ones, which we write as $\varepsilon(J) = \{\varepsilon_j, j \notin \sigma(J)\}$, so that

$$J = \left\{ t \in \mathbb{R}^N : t_j = \varepsilon_j T_j, \text{ if } j \notin \sigma(J), \ 0 \le t_j \le T_j, \text{ if } j \in \sigma(J) \right\}. \quad (4.2.3)$$

Let $\mathcal{O}_k$ denote the $\binom{N}{k}$ elements of $\partial_k T$ which include the origin.

Next, to each sequence $\varepsilon(J)$, define a corresponding set $\varepsilon^*(J)$ of ± 1's, according to the rule $\varepsilon_j^* = 2\varepsilon_j - 1$. Then, with a little rewriting, it follows from the results of Sect. 3.8 that

$$\varphi\left(A_u(f, T)\right) = \sum_{k=0}^{N} \sum_{J \in \partial_k T} \sum_{i=0}^{k} (-1)^i \mu_i(J), \qquad (4.2.4)$$

where, for $i \le \dim(J)$, $\mu_i(J)$ is the number of $t \in J$ for which

$$f(t) \ge u, \qquad (4.2.5)$$

$$f_j(t) = 0, \qquad j \in \sigma(J) \qquad (4.2.6)$$

$$\text{Ind}\left((-f_{mn}(t))_{(m,n \in \sigma(J))} \right) = i, \qquad (4.2.7)$$

$$\varepsilon_j^* f_j(t) > 0, \qquad j \notin \sigma(J) \qquad (4.2.8)$$

and, as usual, the index of a matrix is the number of its negative eigenvalues.

Thus we can find the mean excursion set Euler characteristic by taking expectations of each term on the right of (4.2.4). We start with the term for $k = 0$, temporarily dropping the requirement (4.2.8).

Lemma 4.2.2. *Let f and T be as in Theorem 4.2.1. Set*

$$\mu_k = \#\{t \in T : f(t) \geq u, \ \nabla f(t) = 0, \ \text{Ind}(-\nabla^2 f(t)) = k\}. \qquad (4.2.9)$$

Then

$$\mathbb{E}\left\{\sum_{k=0}^{N}(-1)^k \mu_k\right\} = \frac{|T| \lambda^{N/2}}{(2\pi)^{(N+1)/2}\sigma^N} H_{N-1}\left(\frac{u}{\sigma}\right) e^{-u^2/2\sigma^2}, \qquad (4.2.10)$$

where $|T| = \mathcal{L}_N(T) = \prod_{j=1}^{N} T_j$.

Before turning to the proof of the lemma, there are some crucial points worth noting. The first is that although the definition of the μ_k depends quite strongly on the f_{ij}, the distributions of which involve fourth order spectral moments, these do not appear in the final expectation. As will become clear from the proof, the disappearance of the fourth order spectral moments has a lot to do with the fact that we compute the mean of the alternating sum in (4.2.10) and do not attempt to evaluate the expectations of the individual μ_k. Doing so would indeed involve fourth order spectral moments. The fact that this is all we need is extremely fortunate, for it is actually *impossible* to obtain closed expressions for any of the $\mathbb{E}\{\mu_k\}$.

The second point of interest is that neither (4.2.10) nor any of the results that build on it depend in any way on the long term decay rate of the covariance function of f. It is not hard to see that this is an immediate consequence of the additive property (in the sense of (1.2.2)) of the random variables μ_k.

Proof. To save on notation, we shall assume that $\sigma = \lambda_2 = 1$. The extension to general σ and λ_2 is simple.

Direct application of the Kac-Rice metatheorem, Theorem 4.1.1, applied to each μ_k separately, yields that

$$\mathbb{E}\left\{\sum_{k=0}^{N}(-1)^k \mu_k\right\} = \sum_{k=0}^{N}\int_T (-1)^k \mathbb{E}\left\{\left|\det(\nabla^2 f(t))\right| \mathbb{1}_{\{f(t)\geq u, \text{Ind}(-\nabla^2 f(t))=k\}}\right.$$
$$\left.\left|\nabla f(t) = 0\right\}p_{\nabla f(t)}(0)dt,\right.$$

where

$$p_{\nabla f(t)}(0) = (2\pi)^{-N/2}$$

is the density of $\nabla f(t)$ at 0.

The fact that f is isotropic implies that the pair $(f(t), \nabla^2 f(t))$ is independent of $\nabla f(t)$ (cf.(2.6.6) and (2.6.7)) so we can remove the conditioning in the expectation on the right hand side. This greatly simplifies the calculation.

However, the small miracle that simplifies everything enormously (without which we would not get such a final simple answer) is the fact that

$$(-1)^k \left|\det(\nabla^2 f(t))\right| \mathbf{1}_{\{\mathrm{Ind}(-\nabla^2 f(t))=k\}} = \det(-\nabla^2 f(t)) \mathbf{1}_{\{\mathrm{Ind}(-\nabla^2 f(t))=k\}},$$

which implies that

$$\sum_{k=0}^{N}(-1)^k |\det(\nabla^2 f(t))| \mathbf{1}_{\{\mathrm{Ind}(-\nabla^2 f(t))=k\}} = \det(-\nabla^2 f(t)).$$

Applying this, along with stationarity to integrate out t, then dropping the index t on what remains, we see that

$$\mathbb{E}\left\{\sum_{k=0}^{N}(-1)^k \mu_k\right\} = (2\pi)^{-N/2}|T|\, \mathbb{E}\left\{\det(-\nabla^2 f)\mathbf{1}_{\{f\geq u\}}\right\}$$

$$= (2\pi)^{-N/2}|T|\, \mathbb{E}\left\{\det(-\nabla^2 f - fI + fI)\mathbf{1}_{\{f\geq u\}}\right\}$$

$$= (2\pi)^{-N/2}|T|\, \mathbb{E}\left\{\sum_{j=0}^{N} f^{N-j}\mathrm{detr}_j(-\nabla^2 f - fI)\mathbf{1}_{\{f\geq u\}}\right\},$$

where, as usual, $\mathrm{detr}_j(A)$ is the sum of determinants of all $j \times j$ principle minors of A. Here, we have used the standard expansion that, for an $N \times N$ matrix A,

$$\det(A + \lambda I) = \sum_{j=0}^{N} \lambda^j \mathrm{detr}_j(A).$$

Once again appealing to isotropy, from which it follows that $\mathbb{E}\{f(t)f_{ij}(t)\} = -\delta_{ij}$ (cf. (2.6.5)), it is simple to check that the matrix $\nabla^2 f + fI$ is independent of f. Therefore,

$$\mathbb{E}\left\{\sum_{k=0}^{N}(-1)^k \mu_k\right\} = (2\pi)^{-N/2}|T|\sum_{j=0}^{N}\mathbb{E}\left\{f^{N-j}\mathbf{1}_{\{f\geq u\}}\right\}\mathbb{E}\{\mathrm{detr}_j(-\nabla^2 f - fI)\}.$$

The computation of the expectation of the determinant here is basically algebra, expanding the determinant, applying Wick's formula (2.7.5)–(2.7.6), and again using the fact that $\mathbb{E}\{f(t)f_{ij}(t)\} = -\delta_{ij}$. Doing this[1] leads to

$$\mathbb{E}\left\{\sum_{k=0}^{N}(-1)^k \mu_k\right\} = (2\pi)^{-N/2}|T|\, \mathbb{E}\left\{\sum_{j=0}^{\lfloor\frac{N}{2}\rfloor}\frac{(-1)^j N!}{(N-2j)!j!2^j}f^{N-2j}\mathbf{1}_{\{f\geq u\}}\right\}$$

---

[1] An alternative approach can be based on known results on GUE matrices. See, for example, the computations in Sect. 8.4 of [11], which can easily be adapted to the above situation.

$$= (2\pi)^{-N/2}|T|\mathbb{E}\left\{H_N(f)\mathbb{1}_{\{f\geq u\}}\right\}$$

$$= (2\pi)^{-(N+1)/2}|T|\int_u^\infty H_N(x)\,e^{-x^2/2}\,dx$$

$$= (2\pi)^{-(N+1)/2}|T|H_{N-1}(u)\,e^{-u^2/2}, \tag{4.2.11}$$

where the last line follows directly from the basic properties of Hermite polynomials (cf. (2.2.7)). □

Proof of Theorem 4.2.1. As in the proof of Lemma 4.2.2, we assume that $\sigma = \lambda_2 = 1$. Consider conditions (4.2.5)–(4.2.7). If we restrict f to the face J, then Lemma 4.2.2, modified only to allow for the dimension of J, actually gives the expected number of points satisfying these three conditions. However, we also have to allow for the additional constraint (4.2.8).

To do this, let $\tilde{\mu}_i(J)$ denote the number of points $t \in J$ satisfying (4.2.5)–(4.2.7) while $\mathcal{E}_J(t)$ denotes the event (4.2.8). We need to compute

$$\sum_{i=0}^k (-1)^i \mathbb{E}\left\{\tilde{\mu}_i(J)\mathbb{1}_{\mathcal{E}_J(t)}\right\},$$

and then sum over the faces J of T. However, once again applying isotropy gives us that the random variables in (4.2.8) are independent of those in (4.2.5)–(4.2.7). Therefore, by Lemma 4.2.2, when $k \geq 1$,

$$\sum_{i=0}^k (-1)^i \mathbb{E}\left\{\tilde{\mu}_i(J)\mathbb{1}_{\mathcal{E}_J(t)}\right\} = \frac{|J|}{(2\pi)^{(k+1)/2}}H_{k-1}(u)e^{-u^2/2}\mathbb{P}\{\mathcal{E}_J(t)\}$$

$$= \frac{|J|}{(2\pi)^{(k+1)/2}}H_{k-1}(u)e^{-u^2/2}\frac{1}{2^{N-k}},$$

where the calculation of $\mathbb{P}\{\mathcal{E}_i(t)\}$ follows from symmetry considerations.

It is easy to check from first principles, using the connection between H_{-1} and Ψ, that the above also holds when $k = 0$.

From this and (4.2.4) we finally have

$$\mathbb{E}\left\{\varphi\left(A_u(f,T)\right)\right\} = \sum_{k=0}^N \sum_{J\in\partial_k T} \mathbb{E}\left\{\sum_{i=0}^k (-1)^i\,\tilde{\mu}_i(J)\mathbb{1}_{\mathcal{E}_J}(t)\right\}$$

$$= \sum_{k=0}^N \sum_{J\in\partial_k T} \frac{|J|}{(2\pi)^{(k+1)/2}}H_{k-1}(u)e^{-u^2/2}\frac{1}{2^{N-k}}$$

$$= \sum_{k=0}^N \sum_{J\in\mathcal{O}_k} \frac{|J|}{(2\pi)^{(k+1)/2}}H_{k-1}(u)e^{-u^2/2},$$

where we have used the fact that, for each $J \in \mathcal{O}_k$, there are 2^{N-k} parallel faces of T in $\partial_k T$.

This gives (4.4.4) and so the proof is complete.   □

There is an extension to Theorem 4.2.1, which, while not an immediate corollary of its statement, is easily seen to have an identical proof, modulo notational changes. We shall require it in a moment, and it is given by

Corollary 4.2.3. *Retain the notation and conditions of Theorem 4.2.1, but now let T be an N-dimensional parallelogram. Then the main result, (4.2.1), of the theorem still holds, with the only change being that the Lipschitz-Killing curvatures are now defined accordingly; viz. if T is translated so that the origin 0 is at one of its corners, then*

$$\mathcal{L}_i(T) = \sum_{J: \dim(J)=i,\, 0 \in J} |J|, \tag{4.2.12}$$

where the J are faces of T with volume $|J|$.

4.3 Real Isotropic Fields: Lipschitz-Killing Curvatures

In the previous section we saw how to compute the mean Euler characteristic of the excursion sets of isotropic Gaussian fields on rectangles, and so obtained a very special case of the GKF. The first step towards greater generality lies not in lifting the assumptions on f or the parameter space, but rather on computing mean Lipschitz-Killing curvatures rather than mean Euler characteristics. At this point we consider the Euclidean Lipschitz-Killing curvatures of the Steiner and Weyl formulae, (1.2.7) and (3.4.1).

To do this we use the special case of Crofton's formula known as Hadwiger's formula (cf. (3.7.2)) which we recall is given by

$$\mathcal{L}_k(M) = \int_{\mathrm{Graff}(N,N-k)} \mathcal{L}_0(M \cap V)\, d\lambda^N_{N-k}(V). \tag{4.3.1}$$

where all the terms were defined in Sect. 3.7. The (easy) extension is then as follows:

Theorem 4.3.1. *Let T be a N-dimensional parallelogram, and f such that all the conditions of Theorem 4.2.1 hold. Then, for every $0 \le j \le N$,*

$$\mathbb{E}\left\{\mathcal{L}_j\left(A_u(f,T)\right)\right\} = e^{-u^2/2\sigma^2} \sum_{k=0}^{N-j} \begin{bmatrix} j+k \\ k \end{bmatrix} \frac{\lambda_2^{k/2}}{(2\pi)^{(k+1)/2}\sigma^k} \mathcal{L}_{j+k}(T) H_{k-1}\left(\frac{u}{\sigma}\right),$$

$$\tag{4.3.2}$$

where the $\mathcal{L}_j$, on both sides of the equality, are computed with respect to the standard Euclidean metric on $\mathbb{R}^N$. In particular, on the right hand side they are given by (4.2.2) for rectangles and (4.2.12) for other parallelograms.

Proof. To save on space we shall establish (4.3.2) under the additional assumption that $\sigma^2 = 1$, and also set

$$\rho_j(u) \stackrel{\Delta}{=} (2\pi)^{-(j+1)/2} H_{j-1}(u) e^{-u^2/2}.$$

Then Hadwiger's formula and Corollary 4.2.3 immediately yield

$$\mathbb{E}\left\{\mathcal{L}_j(A_u(f,T))\right\} = \int_{\mathrm{Graff}(N,N-j)} \mathbb{E}\left\{\mathcal{L}_0\left(A_u(f,T) \cap V\right)\right\} d\lambda^N_{N-j}(V)$$

$$= \sum_{k=0}^{N-j} \lambda_2^{k/2} \rho_k(u) \int_{\mathrm{Graff}(N,N-k)} \mathcal{L}_k(T \cap V) d\lambda^N_{N-j}(V)$$

$$= \sum_{k=0}^{N-j} \begin{bmatrix} j+k \\ k \end{bmatrix} \lambda_2^{k/2} \rho_k(u) \mathcal{L}_{j+k}(T),$$

and we are done. □

For an alternative approach to computing the mean Lipschitz-Killing curvatures in the setting of this section see [92].

4.4 Real Stationary Fields on Rectangles: Euler Characteristic

In this section we shall see how to extend Theorem 4.2.1 from isotropic Gaussian fields over rectangles to stationary fields. This is still not the full GKF, but we are getting closer. Unfortunately, it is the last case that we can treat without getting into differential geometry and curvature integrals.

Consider, therefore, f to be Gaussian, stationary, zero mean, variance σ^2 and satisfying the smoothness conditions of Theorem 4.2.1 over the rectangle $T = \prod_{j=1}^N [0, T_j]$. Since f is no longer isotropic, the second spectral moments

$$\lambda_{ij} = \mathbb{E}\{f_i(t) f_j(t)\} \tag{4.4.1}$$

will not be as simple as in the isotropic case, and we denote the $N \times N$ matrix of these moments by Λ. Recall that in the isotropic case we had $\Lambda = \lambda_2 I$, a fact which simplified many of the computations in Sect. 4.2.

Let Q be a positive definite square root of Λ^{-1}, so that

$$Q'\Lambda Q = I. \tag{4.4.2}$$

Note that $\det Q = (\det \Lambda)^{-1/2}$. Now take the transformation of $\mathbb{R}^N$ given by $t \to tQ^{-1}$, under which the rectangle T transforms to the parallelogram $T^Q = \{\tau : \tau = tQ^{-1} \text{ for some } t \in T\}$ and define $f^Q : T^Q \to \mathbb{R}$ by

$$f^Q(t) \triangleq f(tQ).$$

The new process f^Q has covariance function

$$C^Q(s,t) = C(sQ, tQ) = C((t-s)Q)$$

and so is still stationary, with constant variance σ^2. Furthermore, simple differentiation shows that $\nabla f^Q = (\nabla f)Q$, from which it follows that

$$\begin{aligned} \Lambda^Q &\triangleq \mathbb{E}\left\{((\nabla f^Q)(t))'((\nabla f^Q)(t))\right\} \\ &= Q'\Lambda Q \\ &= I. \end{aligned} \tag{4.4.3}$$

That is, the first order derivatives of the transformed process are now uncorrelated and of unit variance. We now show that it is sufficient to work with this, much simpler, transformed process.

Firstly, we note the crucial fact that the μ_k of (4.2.9) for f over T are *identical* to those for f^Q over T^Q. Clearly, there is a trivial one-one correspondence between those points of T at which $f(t) \geq u$ and those of T^Q at which $f^Q(t) \geq u$. We do, however, need to check more carefully what happens with the conditions on ∇f and $\nabla^2 f$.

Since $\nabla f^Q = (\nabla f)Q$, we have that $(\nabla f^Q)(t) = 0$ if, and only if, $\nabla f(tQ) = 0$. In other words, there is also a simple one-one correspondence between critical points. Furthermore, since $\nabla^2 f^Q = Q'(\nabla^2 f)Q$ and Q is a positive definite matrix, $\nabla^2 f^Q(t)$ and $\nabla^2 f(tQ)$ have the same index.

Consequently, we can now work with f^Q rather than f. While f^Q is not isotropic, it is 'locally isotropic' in the sense of Sect. 2.6.2. More importantly, if you look back through the proof of Corollary 4.2.1, you will find that *nowhere in the proof did we use the full force of isotropy,[2] but only local isotropy!* Thus Theorem 4.2.1 can be applied to f^Q, as can Theorem 4.2.3.

All that remains, then, to apply Corollary 4.2.3 is to compute the LKCs of the parallelogram T^Q. Using Steiner's Formula (1.2.7) it is easy geometry

---

[2] On the other hand, this is not true of Theorem 4.3.1 where we did need full isotropy to apply Hadwiger's Theorem.

to see that these are given by

$$\mathcal{L}_k(T^Q) = \sum_{J \in \mathcal{O}_k} |J| |\det \Lambda_J|^{1/2},$$

where the sum is over all the k-dimensional faces of T^Q in $\mathcal{O}_k$, itself the $\binom{N}{k}$ elements of $\partial_k T$ which include the origin. As for the matrix Λ_J, recall that each k-dimensional face $J \in \partial_k T$ has a representation as in (4.2.3) via a subset $\sigma(J)$ of $\{1, \ldots, N\}$. Using this representation, we write Λ_J for the $k \times k$ matrix with elements λ_{ij}, $i, j \in \sigma(J)$. The volume $|J| = \prod_{i \in \sigma(J)} T_i$ is the usual k-dimensional measure of the face J.

The following result is now an immediate consequence of the above and Corollary 4.2.3.

Theorem 4.4.1. *Let f be stationary rather than isotropic, and otherwise assume that all the conditions of Theorem 4.2.1 hold. Then*

$$\mathbb{E}\left\{\varphi\left(A_u(f, T)\right)\right\} = e^{-u^2/2\sigma^2} \sum_{k=1}^{N} \sum_{J \in \mathcal{O}_k} \frac{|J| |\Lambda_J|^{1/2}}{(2\pi)^{(k+1)/2} \sigma^k} H_{k-1}\left(\frac{u}{\sigma}\right) + \Psi\left(\frac{u}{\sigma}\right).$$

$$(4.4.4)$$

4.5 The Induced Metric and the Need for Riemannian Geometry

The flow of the proofs so far in this chapter has been as follows:

(1) Use Morse theory to express the Euler characteristic of excursion sets via alternating sums of numbers of critical points of different index.
(2) Use the Kac-Rice metatheorem to find a (complicated, integral) expression for the expectation.
(3) Show that in the (locally) isotropic case the integral can be evaluated for fields over rectangles and parallelograms.
(4) Use Hadwiger's formula to go from the expected Euler characteristic to expected Lipschitz-Killing curvatures.
(5) Use a simple transformation to turn the Euler characteristic computation for stationary fields on rectangles into problem about locally isotropic fields on parallelograms.

To reach the level of generality that we really want for the Gaussian kinematic formula, we need to do the following:

(a) Weaken the assumption of stationarity to one of constant variance.
(b) Move from rectangles to general parameter spaces.
(c) Move from real valued fields to vector valued ones.

It turns out that Step (a) involves the main conceptual leap, after which (b) and (c) are (not small) strides forward. The trick in solving (a) is to find a way to turn our non-isotropic, non-stationary random fields, which may now also be defined on a stratified manifold, into locally isotropic ones. Local isotropy, after all, was the key to permitting explicit computation in (3).

As already noted in Sect. 2.6.2, even on $\mathbb{R}^N$ there is no simple transformation to either isotropy or even local isotropy for general non-stationary fields. The basic problem is that different transformations may be required in different parts of the parameter space, and patching them together to make something which is globally well defined may not be easy. In fact, it is *not* easy, but under the assumption of constant variance there is a technique for doing it that works well for our needs, and this involves replacing the usual Euclidean inner product between vectors by an appropriate Riemannian metric. Thus, while we do not transform the parameter space at all, *we change the way we measure things on it.*

As we discussed briefly in Sect. 3.1, a Riemannian metric g on a manifold M is a family, $\{g_t\}_{t \in M}$, of inner products on the tangent spaces $T_t M$. Keeping M a nice set within $\mathbb{R}^N$ for the moment, it suffices to define the metric on unit vectors parallel to the axes. Thus, if $t \in M$ and we define the vector based at t and parallel to the j-th axis as

$$E_j(t) = t + e_j,$$

where, as usual, e_j is the vector with 1 in the j-th position and 0 elsewhere, then a Riemannian metric which is suited to our purposes is defined on the E_j by

$$g_t\left(E_i(t),\, E_j(t)\right) = (\sigma^{-2}\Lambda(t))_{ij} \stackrel{\Delta}{=} \sigma^{-2}\,\mathrm{Cov}\left(\frac{\partial f(t)}{\partial t_i},\, \frac{\partial f(t)}{\partial t_j}\right)$$

$$\equiv \sigma^{-2}\mathbb{E}\left\{E_i(t)f(t) \cdot E_j(t)f(t)\right\}, \quad (4.5.1)$$

and extended to other vectors by linearity. We call g_t the *metric induced by the random field f.* Note that to define this metric we have assumed that f has constant variance σ^2. However, neither stationarity nor isotropy is needed for the definition to make sense.

In the isotropic and locally isotropic cases, the induced metric is the usual Euclidean metric, multiplied by a factor of λ_2, the second spectral moment. In the stationary case,

$$g_t(U, V) = \langle \sigma^{-1}U\Lambda^{1/2},\, \sigma^{-1}V\Lambda^{1/2} \rangle = \langle U,\, \sigma^{-2}V\Lambda \rangle,$$

where the right hand inner products are the usual Euclidean one and Λ is the matrix (4.4.1) of second order spectral moments. In both the stationary and isotropic cases g_t is independent of t.

Note that the induced metric is scale independent, since if we multiply f by σ then both the variance of f and the covariances of its derivatives will change by a factor of σ^2 leaving g is unchanged. Thus g is purely a measure of the spatial variation of f.

In the case of a random field defined on a general manifold M, we similarly define the induced metric by

$$g_t\left(X_t, Y_t\right) \stackrel{\Delta}{=} \mathbb{E}\left\{X_t f(t) \cdot Y_t f(t)\right\}, \qquad X_t, Y_t \in T_t M.$$

Recall now from Sect. 3.1 that Riemannian metrics determine a notion of volume. As a consequence, given a Riemannian metric and a nice enough set M, it is also possible to define Lipschitz-Killing curvatures $\mathcal{L}_j(M)$ which correspond to the metric. All one needs to do is take the original definition (3.4.11) of Lipschitz-Killing curvatures, and interpret the curvature tensor, $\widetilde{R}$, the second fundamental form $\widetilde{S}$ and Hausdorff measures $\mathcal{H}$ as Riemannian objects, induced on the manifold and its tangent bundle by the metric. While our original approach to Lipschitz-Killing curvatures came from tube formulae, there is no reason to believe that, on a general Riemannian manifold, a tube formula involving these new, Riemannian, Lipschitz-Killing curvatures will hold. Nevertheless, there is no problem with defining them.

For the full theory, and interesting examples of computation of the Lipschitz-Killing curvatures, you should go to either *RFG* or *ARFG*. But we have seen some examples already.

For example, when looking at the supremum distribution of the cosine random field in Sect. 2.2 we came up with the expression (cf. (2.2.2))

$$\sum_{j_1 \cdots j_k} \sigma^{-k} \prod_{i=1}^{k} \lambda_{j_i} T_{j_i}. \tag{4.5.2}$$

This is precisely $\mathcal{L}_k(T)$ (T is the usual rectangle) where $\mathcal{L}_k$ is calculated with respect to the metric induced by the cosine field. In general, (4.5.2) is appropriate for a random field with a matrix of second order spectral moments of the form $\operatorname{diag}(\lambda_{jj}) = \operatorname{diag}(\lambda_j^2)$.

Similarly, in Theorem 4.4.1 we met the expression

$$\sigma^{-k} \sum_{J \in \mathcal{O}_k} |J| \, |\Lambda_J|^{1/2}, \tag{4.5.3}$$

(cf. (4.4.4).) But it is easy to check from the above descriptions that this is, once again, precisely $\mathcal{L}_k(T)$, calculated with respect to the metric induced by the stationary random field of the theorem.

Now recall (3.5.3), which connected Hermite polynomials with Gaussian Minkowski functionals, and we can rewrite Theorem 4.4.1 as follows.

Theorem 4.5.1. *Let T be an N-dimensional parallelogram, and assume that f satisfies the conditions of Theorem 4.2.1 assuming stationarity rather than isotropy and variance $\sigma^2 = 1$. Then*

$$\mathbb{E}\left\{\varphi\left(A_u(f,T)\right)\right\} = \sum_{j=0}^{\dim T} (2\pi)^{-j/2}\mathcal{L}_j(T)\mathcal{M}_j([u,\infty)), \qquad (4.5.4)$$

where the Lipschitz-Killing curvatures $\mathcal{L}_k(M)$ are computed with respect to the Riemannian metric induced on T by f.

Actually, with Theorem 4.5.1 established, it does not require a lot of imagination, or faith, to believe that what is true for parallelograms T must also be true for regular stratified manifolds M. Indeed, it is true, and even without the condition of stationarity,[3] although it takes some work to prove it. However, before moving to more general parameter spaces, we need an appropriate definition of regularity for random fields over stratified manifolds. This is given by

Condition 4.5.2 (Regularity of a random field). *Let M be a regular stratified manifold, and f be a centered Gaussian field over M. Let $\mathcal{A} = (U_\alpha, \varphi_\alpha)_{\alpha \in I}$ be a countable atlas for M. Then we say that f is regular if, for every α, the Gaussian field $f_\alpha = f \circ \varphi_\alpha^{-1}$ on $\varphi_\alpha(U_a) \subset \mathbb{R}^N$ satisfies conditions (i) and (ii) of Theorem 4.2.1 for each $T = \varphi_\alpha(U_\alpha)$, $f = f_\alpha$ and some $K = K_\alpha > 0$.*

Note that if M is a rectangle, then the definition of regularity reduces to the conditions in Theorem 4.2.1. Among other things, Condition 4.5.2 ensures that the sample functions of f are, with probability one, Morse functions over M.

Theorem 4.5.1, with the parallelograms T replaced by regular stratified manifolds, is a far more general special case of the Gaussian kinematic formula than any of the preceeding theorems of this chapter. In particular, we have finally seen why one needs the induced Riemannian metric.

However, we still have steps (b) and (c) above to confront in order to move to full generality. This is the material of the following, final, sections of this chapter.

---

[3] This is good, since 'stationarity' is not easy to define on a general manifold, for which one would need there to be a group action on M with regard to which the distribution of f is shift invariant. Obviously, not all manifolds possess such a group structure.

4.6 The Canonical Isotropic Process on the Sphere

In this section we shall show how to tackle (b) and (c) of the preceeding section, by indicating how to prove a version of the Gaussian kinematic formula for a vector valued random field on nice subsets of the sphere. The main restriction will be that we shall work with only a very special process, the canonical, isotropic Gaussian process on the sphere. However, we have already seen in Sect. 2.4 that this process provides a model for many others, a point that we shall return to later in this chapter when we look at more general results.

Recall from Sect. 2.4 that the real valued canonical isotropic Gaussian process on the l-sphere $S(\mathbb{R}^l)$ is defined as the centered Gaussian process on $S(\mathbb{R}^l)$ with covariance function

$$\mathbb{E}\left\{f(s)f(t)\right\} = \langle s, t \rangle, \tag{4.6.1}$$

where $\langle\,,\,\rangle$ is the usual Euclidean inner product. This process can be realised as

$$f(t) = \langle \xi,\, t \rangle,$$

where $\xi \sim N(0, I_{k \times k})$. We shall be interested in a k-dimensional version of this process, which we still denote by f, for which the real valued coordinate processes are independent centered processes satisfying (4.6.1). We care about the excursion set

$$A_D = A_D(f, M) = M \cap f^{-1}(D),$$

where both M and D are regular stratified manifolds, $M \subset S(\mathbb{R}^l)$ and $D \subset \mathbb{R}^k$. Our aim now is to try to establish the full Gaussian kinematic formula (4.0.1) in this setting.

Our approach will be a little circuitous, but along the way we shall see that the Gaussian kinematic formula is closely related to the kinematic fundamental formula (1.3.9) from which relationship comes its name.

4.6.1 A Model Process

We start with a family, $\{y^{(n)})\}_{n \geq l}$, of smooth $\mathbb{R}^k$ valued processes on $S(\mathbb{R}^l)$. The limit of these processes will be the canonical Gaussian processes, but, as we shall soon see, they are simpler to handle than their limit. To define the family, for each $n \geq l$ we first embed $S(\mathbb{R}^l)$ in $S(\mathbb{R}^n)$ in the natural way, by setting

$$S(\mathbb{R}^l) = \{t = (t_1, \ldots, t_n) \in S(\mathbb{R}^n) : t_{l+1} = \cdots = t_n = 0\}.$$

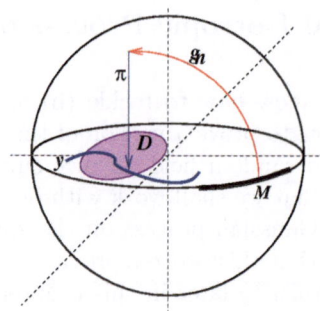

Fig. 4.6.1 The pre-Gaussian process $y^{(2)}$ from a (one-dimensional) subset of a great circle to $\mathbb{R}^2$

We next take the rotation group, $O(n)$, thinking of it as the group of orthonormal $n \times n$ matrices, equipped with its normalized Haar measure μ_n, as our underlying probability space. Then the n-th process $y^{(n)}$ is defined by

$$y^{(n)}(t, g_n) \triangleq \pi_{n,k}\left(\sqrt{n}g_n t\right), \qquad (4.6.2)$$

where $t \in S(\mathbb{R}^l)$, $g_n \in O(n)$ and $\pi_{n,k}$ is the projection from $S_{\sqrt{n}}(\mathbb{R}^n)$ to $\mathbb{R}^k$ given by

$$\pi_{n,k}(x_1, \ldots, x_n) = (x_1, \ldots, x_k), \qquad (4.6.3)$$

An example is given in Fig. 4.6.1.

To see why the $y^{(n)}$ should converge to the $\mathbb{R}^k$ valued canonical Gaussian process on $\mathbb{R}^l$, we need a version of what is generally known as the Poincaré limit theorem (cf. [29] for some history) as well as a more recent, more powerful version due to Diaconis, Eaton and Lauritzen [30].

The classic Poincaré limit theorem states that if $\eta_n = (\eta_{n1}, \ldots, \eta_{nn})$ is uniformly distributed on $S_{\sqrt{n}}(\mathbb{R}^n)$ and $k \geq 1$ is fixed, then the joint distribution of $(\eta_{n1}, \ldots, \eta_{nk})$ converges weakly to that of k independent standard Gaussians as $n \to \infty$. That is, if

$$X_{k,n} \triangleq \pi_{n,k}(\eta_n),$$

then, as $n \to \infty$, and for k fixed

$$X_{k,n} \overset{\mathcal{L}}{\to} N(0, I_{k \times k}), \qquad (4.6.4)$$

where $\overset{\mathcal{L}}{\to}$ denotes convergence in distribution.

This can be proven either by realising η as a vector of standard normal random variables conditioned to lie on the sphere, or via a direct calculus argument.

Convergence in distribution can be lifted to total variation convergence, and from there to the convergence of the finite dimensional distributions of

$y^{(n)}$ to those of f. More importantly, for functionals F of these processes for which $\mathbb{E}\{|F(y)|\} < \infty$, we have [29, 30]

$$\lim_{n \to \infty} \mathbb{E}\{F(y^{(n)})\} = \mathbb{E}\{F(y)\}. \tag{4.6.5}$$

As we are about to see, it is remarkably straightforward to compute the mean Lipschitz-Killing curvatures of excursion sets of the $y^{(n)}$, using only the kinematic fundamental formula on $S_{\sqrt{n}}(\mathbb{R}^n)$. This is the content of the following subsection. We shall then wave our hands to send $n \to \infty$ to 'complete' the proof for the canonical Gaussian process.

4.6.2 The GKF for the Model Process

Of course, there is nothing Gaussian about the model processes $y^{(n)}$, so the G in the GKF of the section title is a little misleading. Nevertheless, we shall try for a parallel result, and thus obtain a key lemma of major importance. The lemma is the following, under the usual conditions on M and D.

Lemma 4.6.1. *Let $y^{(n)}$ be the model process* (4.6.2) *on $M \subset S(\mathbb{R}^l)$, with $n \geq l$. Then,[4] for $D \subset \mathbb{R}^k$,*

$$\mathbb{E}\left\{\mathcal{L}_i^1(M \cap (y^{(n)})^{-1}D)\right\}$$

$$= \sum_{j=0}^{\dim M - i} \left(n^{j/2} \begin{bmatrix} n-1 \\ j \end{bmatrix}^{-1}\right) \begin{bmatrix} i+j \\ j \end{bmatrix} \mathcal{L}_{j+i}^1(M) \frac{\mathcal{L}_{n-1-j}^{n-1}\left(\pi_{n,k}^{-1}(D)\right)}{s_n n^{(n-1)/2}}$$

$$= \sum_{j=0}^{\dim M - i} \frac{s_{i+1}}{s_{i+j+1}} \mathcal{L}_{j+i}^1(M) \frac{\mathcal{L}_{n-1-j}^{n-1}\left(\pi_{n,k}^{-1}(D)\right)}{s_{n-j} n^{(n-1-j)/2}}, \tag{4.6.6}$$

where

$$s_n = \frac{2\pi^{n/2}}{\Gamma(\frac{n}{2})} \tag{4.6.7}$$

is the surface area of the unit ball in $\mathbb{R}^n$.

---

[4] The meaning of $\pi_{n,k}^{-1}(D)$ in (4.6.6) is a little subtle. The problem is that, for all $t \in S_{\sqrt{n}}(\mathbb{R}^n)$, $\pi_{n,k}(t) \in B_{\sqrt{n}}(\mathbb{R}^n)$, which may, or may not, cover D. Thus, since

$$\pi_{n,k}^{-1}(D) = \left\{t \in S_{\sqrt{n}}(\mathbb{R}^n) : \pi_{n,k}(t) \in D\right\},$$

it follows that $\pi_{n,k}^{-1}(D)$ may be only the inverse image of a subset of D. However, as long as D has finite diameter, this problem will disappear for n large enough.

Proof. Since $\pi_{n,k}^{-1}(D)$ is a nice domain in $S(\mathbb{R}^n)$, it follows from the construction of $y^{(n)}$ (see Fig. 4.6.1) that

$$
\begin{aligned}
& \mathbb{E}\left\{\mathcal{L}_i^1(M \cap (y^{(n)})^{-1}D)\right\} \\
&= \int_{O(n)} \mathcal{L}_i^1(M \cap (y^{(n)})^{-1}D)(g_n)\,d\mu_n(g_n) \\
&= \int_{O(n)} \mathcal{L}_i^1\left(M \cap n^{-1/2}g_n^{-1}\left(\pi_{n,k}^{-1}(D)\right)\right)\,d\mu_n(g_n) \\
&= n^{-i/2}\int_{O(n)} \mathcal{L}_i^{n-1}\left(\sqrt{n}M \cap g_n^{-1}\left(\pi_{n,k}^{-1}(D)\right)\right)\,d\mu_n(g_n) \\
&= \frac{1}{s_n n^{(n-1+i)/2}}\int_{G_{n,n-1}} \mathcal{L}_i^{n-1}\left(\sqrt{n}\,M \cap g_n\left(\pi_{n,k}^{-1}(D)\right)\right)\,d\nu_{n,n-1}(g_n),
\end{aligned}
$$

where the second last line follows from the scaling properties of Lipschitz-Killing curvatures (e.g. (1.2.13)) and the last is really no more than a notational change, using the definition (3.6.1) of $\nu_{n,\lambda}$.

However, applying the kinematic fundamental formula (3.6.2) to the last line above, we immediately have that it is equal to

$$
\begin{aligned}
& \sum_{j=0}^{\dim M - i} n^{j/2}\begin{bmatrix} i+j \\ i \end{bmatrix}\begin{bmatrix} n-1 \\ j \end{bmatrix}^{-1} \frac{\mathcal{L}_{j+i}^{n-1}\left(\sqrt{n}\,M\right)}{n^{(i+j)/2}}\frac{\mathcal{L}_{n-1-j}^{n-1}\left(\pi_{n,k}^{-1}(D)\right)}{s_n n^{(n-1)/2}} \\
&= \sum_{j=0}^{\dim M - i} n^{j/2}\begin{bmatrix} n-1 \\ j \end{bmatrix}^{-1}\begin{bmatrix} i+j \\ j \end{bmatrix}\mathcal{L}_{j+i}^1(M)\frac{\mathcal{L}_{n-1-j}^{n-1}\left(\pi_{n,k}^{-1}(D)\right)}{s_n n^{(n-1)/2}},
\end{aligned}
$$

which proves the lemma.   □

Lemma 4.6.1 is starting to take the form of the Gaussian kinematic formula (4.0.1). The combinatorial flag coefficients are in place, as are curvatures. Admittedly, we have $\mathcal{L}_{j+i}^1$ curvatures rather than the $\mathcal{L}_{j+i}$, but since our parameter space is part of a sphere that is to be expected. The important fact is that on the right hand side of the equation we have already managed a split into product form, each factor of which depends on the underlying manifold M or the set D, but not both.

We cannot over-emphasise how important and yet how simple is Lemma 4.6.1. The simplicity lies in the fact that, despite the preponderence of subscripts and superscripts, the proof uses no more than the kinematic fundamental formula and a few algebraic manipulations. Of course, the price that we have paid for this simplicity is that we have a result only for a very special process, and not for the Gaussian processes of central interest to us. However, these are obtainable as appropriate limits of the $y^{(n)}$, as we shall now see.

4.6.3 Back to the Canonical Process

Suppose we send $n \to \infty$ in (4.6.6), which by Poincaré's limit is effectively equivalent to replacing the model process $y^{(n)}$ with a $\mathbb{R}^k$ valued canonical Gaussian f. Then, in order for $\mathbb{E}\{\mathcal{L}_j(M \cap f^{-1}D)\}$ to be finite for the limit process f, we would like to have the following limits existing for each $j < \infty$:

$$\widetilde{\rho}_j(D) \overset{\Delta}{=} \lim_{n\to\infty} n^{j/2} \begin{bmatrix} n-1 \\ j \end{bmatrix}^{-1} \frac{\mathcal{L}_{n-1-j}^{n-1}\left(\pi_{n,k}^{-1}(D)\right)}{s_n n^{(n-1)/2}}. \tag{4.6.8}$$

A Stirling's formula computation shows that if the limit here exists, then

$$\widetilde{\rho}_j(D) = (2\pi)^{-j/2}[j]! \lim_{n\to\infty} \frac{\mathcal{L}_{n-1-j}^{n-1}\left(\pi_{n,k}^{-1}(D)\right)}{s_n n^{(n-1)/2}}. \tag{4.6.9}$$

Sending $n \to \infty$ in Lemma 4.6.1 and applying Poincaré's limit (4.6.5), we see that, if $\mathbb{E}\left\{|\mathcal{L}_i^1(M \cap f^{-1}D)|\right\} < \infty$, then

$$\mathbb{E}\left\{\mathcal{L}_i^1(M \cap f^{-1}D)\right\} = \lim_{n\to\infty} \mathbb{E}\left\{\mathcal{L}_i^1(M \cap (y^{(n)})^{-1}D)\right\}$$

$$= \sum_{j=0}^{\dim M - i} \begin{bmatrix} i+j \\ i \end{bmatrix} \mathcal{L}_{j+i}^1(M)\widetilde{\rho}_j(D). \tag{4.6.10}$$

This is almost (4.0.1), the result that we are trying to prove. Again, the combinatorial flag coefficients are in place and the move from the $\mathcal{L}_{j+i}^1$ to the $\mathcal{L}_{j+i}$ is easily accomplished via (3.4.23) and (3.4.24). Thus, at the cost of perhaps changing the definitions of the $\widetilde{\rho}_j(D)$, we could drop the superscript 1 on both sides of (4.6.10).

All that remains to show is that (new) $\widetilde{\rho}_j(D)$ are equal to $(2\pi)^{-j/2}\mathcal{M}_j^\gamma(D)$, and we shall have

Theorem 4.6.2. *Let $f : S(\mathbb{R}^l) \to \mathbb{R}^k$ be the k-dimensional canonical Gaussian process, and $M \subset S(\mathbb{R}^l)$ and $D \subset \mathbb{R}^k$ be regular stratified manifolds. Then*

$$\mathbb{E}\{\mathcal{L}_i(A_D(f,M))\} = \sum_{j=0}^{\dim M - i} \begin{bmatrix} i+j \\ j \end{bmatrix} (2\pi)^{-j/2}\mathcal{L}_{i+j}(M)\mathcal{M}_j^\gamma(D). \tag{4.6.11}$$

Unfortunately, showing that $\widetilde{\rho}_j(D) = (2\pi)^{-j/2}\mathcal{M}_j^\gamma(D)$ is a rather long and tricky task, involving some delicate (Riemannian) spherical geometry. You can find all the details in *RFG*.

4.7 Fields with Finite Expansions

Recall once more the arguments of Sect. 2.4, where we saw that virtually any
unit variance Gaussian process f with an orthonormal expansion of order
$l < \infty$ can be realised as the canonical isotropic process, which we denoted
there and will denote here as $\widetilde{f}$, on an appropriately chosen subset of $S(\mathbb{R}^l)$.
In fact, if f is defined over a regular stratified manifold M, then the corresponding subset of $S(\mathbb{R}^l)$, which we denote by $\widetilde{\varphi}(M)$ for consistency with the
notation of Sect. 2.4, is also a regular stratified manifold. The relationship
was

$$f(t) = \widetilde{f}\,(\widetilde{\varphi}(t)), \qquad t \in M,\ \widetilde{\varphi}(t) \in \widetilde{\varphi}(M) \subset S(\mathbb{R}^l) \qquad (4.7.1)$$

In Sect. 2.4 we used this fact to study the distribution of the supremum
of f on M by relating it to that of $\widetilde{f}$ on $\widetilde{\varphi}(M)$. It is natural to try the same
approach for the Gaussian kinematic formula, and, in fact, it works.

Note first that the excursion sets of f and $\widetilde{f}$ are related by the fact that
$A_D(\widetilde{f}, \widetilde{\varphi}(M)) = \widetilde{\varphi}(A_D(f, M))$ and so, as long as $\widetilde{\varphi}$ is smooth enough (C^2 and
1–1 suffices) it is easy to relate the Lipschitz-Killing curvatures of $A_D(f, M)$
to those of $A_D(\widetilde{f}, \widetilde{\varphi}(M))$. Their Euler characteristics, for example, will be
identical. As far as the others are concerned, the Lipschitz-Killing curvatures
of a set $A \in \widetilde{\varphi}(M)$, computed with respect to the usual Euclidean metric
on $S(\mathbb{R}^l)$, will be identical to those of $\widetilde{\varphi}^{-1}(A) \in M$ computed with respect
to the Riemannian metric on M which is the pull-back by $\widetilde{\varphi}^{-1}$ to M of the
Euclidean metric on $\widetilde{\varphi}(M)$. Since another calculation like (2.4.2) shows that
this is precisely the metric induced on M by the process f we have that, if
$A \subset M$, then

$$\mathcal{L}_j^f(A) \equiv \mathcal{L}_j^{\widetilde{f}}(\widetilde{\varphi}(A)),$$

where superscripts have been added to emphasise that the Lipschitz-Killing
curvatures on the left are computed with respect to the metric induced by
f on M, while on the right they are computed with respect to the metric
corresponding to $\widetilde{f}$ on $S(\mathbb{R}^l)$; viz. the usual Euclidean metric.

These observations, along with Theorem 4.6.2 actually suffice to establish
the full Gaussian kinematic formula of the following section for all processes satisfying the assumptions of Theorem 4.8.1 which also have a finite
expansion.

The natural question to ask now is therefore whether or not we can extend
this approach to processes without a finite expansion, thereby avoiding all
the Morse theoretic computations which are needed in RFG to establish the
general result. The only truly honest answer we have to this is "We have not
been able to do so, although we have tried". Also true is the claim that "We
doubt it is possible". But doubts such as these are often ephemeral.

4.8 The GKF in the General Case

We are now ready to state the Gaussian kinematic formula for the most general case for which we know it to be true.

Theorem 4.8.1. *Let $M \subset \mathbb{R}^N$ and $D \subset \mathbb{R}^k$ be regular stratified manifolds. Let $f = (f_1, \ldots, f_k) : M \to \mathbb{R}^k$ be a vector valued random process, the components of which are independent, identically distributed, real valued, Gaussian processes, regular in the sense of Condition 4.5.2, and with zero mean and constant unit variance. Then*

$$\mathbb{E}\left\{ \mathcal{L}_i \left(M \cap f^{-1}(D) \right) \right\} = \sum_{j=0}^{\dim M - i} \begin{bmatrix} i+j \\ j \end{bmatrix} (2\pi)^{-j/2} \mathcal{L}_{i+j}(M) \mathcal{M}_j^\gamma(D), \quad (4.8.1)$$

where the $\mathcal{L}_j$, $j = 0, \ldots, N$ are the Lipschitz-Killing measures of M with respect to the metric induced by the f_i, and the $\mathcal{M}_j^\gamma$ are the Gaussian Minkowski functionals on $\mathbb{R}^k$.

We are not going to try to prove Theorem 4.8.1 here. The proof, however, relies on an old technique from integral geometry that, in the context of our problem, argues as follows: We want to prove Theorem 4.8.1, in particular, (4.8.1). Suppose we could show that there exist functions ρ_{ij} on nice sets $D \in \mathbb{R}^k$ independent of the distribution of f and the topology of M, such that

$$\mathbb{E}\left\{ \mathcal{L}_i(M \cap f^{-1}(D)) \right\} = \sum_{j=0}^{\dim M - i} \mathcal{L}_{i+j}(M) \, \rho_{ij}(D). \quad (4.8.2)$$

Then, in order to identify the function ρ_{ij}, we could choose a parameter space and random process that were simple enough to enable us to compute $\mathbb{E}\left\{ \mathcal{L}_i(M \cap f^{-1}(D)) \right\}$ in full. Writing it in the form of (4.8.2) would then allow us to determine the ρ_{ij}, and so we would have the result in full generality. In fact, this is precisely what we have done, by working with the canonical Gaussian process on the sphere.

All that remains, therefore, for us to have a full proof of Theorem 4.8.1 is to prove (4.8.2). The proof of this starts along the lines of that for Theorem 4.2.1, where we looked at isotropic fields on rectangles. That is, is starts with some Morse theory and an attempt to compute the mean value of the Euler characteristic. The detailed computations very rapidly become extremely complicated, although one can get them to the point that (4.8.2) appears, albeit only with $i = 0$; viz. only for the Euler characteristic. To move from the Euler characteristic to general Lipschitz-Killing curvatures one needs an argument akin to the Crofton/Hadwiger formula argument of Sect. 4.3, which we applied to a similar end in the Euclidean, isotropic case. Now, however, we no longer have isotropy, nor is the setting necessarily

Euclidean. Thus, in *RFG* a new kind of Crofton formula is developed, for Riemannian manifolds, which solves the problem. This gives (4.8.2) for all Lipschitz-Killing curvatures, and we are done.

For later reference, here is an immediate corollary of Theorem 4.8.1.

Corollary 4.8.2. *Under the conditions of Theorem 4.8.1, with $k = 1$,*

$$\mathbb{E}\left\{\varphi\left(A_u(f, M)\right)\right\} = e^{-u^2/2} \sum_{j=0}^{\dim M} (2\pi)^{-(j+1)/2} \mathcal{L}_j(M) H_{j-1}(u). \qquad (4.8.3)$$

4.9 Not Just Excursion Sets

Before we leave the theory of the Gaussian kinematic formula for applications, there is one, rather important, aspect to it that deserves noting, and is easily missed at first reading. Recall that in Theorem 4.8.1 f is a (random) mapping of a N-dimensional parameter set into a k-dimensional parameter space. We never said much about the dimension of the set D, but note that there is no need for it to be of full dimension in $\mathbb{R}^k$. If $\dim(D) \leq k$, then, since f is smooth,

$$\dim\left(M \cap f^{-1}D\right) = \dim(M) + k - \dim(D),$$

with probability one. Thus, for example, if $\dim(D) < k$, then $M \cap f^{-1}D$ will be a submanifold of M, and so the information given by its Lipschitz-Killing curvatures is of particular interest.

The case which we have discussed in most detail is that for which $k = 1$ and D is a semi-infinite interval $[u, \infty)$. Then the Gaussian kinematic formula gives information about the mean geometry of excursion sets. However, it can also be applied directly to the boundary ∂D rather than D itself, giving information about the boundary of the excursion set. This kind of problem has been studied in some detail by Wschebor and others (e.g. [93]) with techniques somewhat different to those used to study the excursion sets themselves. The Gaussian kinematic formula, however, includes all these scenarios in a single, unified, result.

4.10 Infinite Dimensions

As these words are being written, Jonathan Taylor and Sreekar Vadlamani are dotting the i's and crossing the t's in the proofs in an important paper [82] that gives an infinite dimensional variant of the Gaussian kinematic formula.

Their result, which involves a heavy dose of stochastic analysis à la Malliavin, is too technical to even quote here, without developing a considerable amount of background and notation. Nevertheless, it is worthwhile spending a little of the publisher's ink and investing a little of your attention to understand what a result like this might look like, and why it is important.

The Gaussian kinematic formula treated Gaussian processes from a finite dimensional parameter space M to $\mathbb{R}^k$. However, there are many examples in both pure and applied probability in which it is natural to take $k = \infty$. For example, to each point $t \in M$ we might associate the entire realisation of another Gaussian process, say $\{f_t(s)\}_{s \in M'}$. A natural case would be rewriting a space-time process $f(t, s)$, $t \in M$, $s \in [a, b]$ as

$$f_t(s) \stackrel{\Delta}{=} f(t, s),$$

so that the path at the point t is $\{f_t(s)\}_{s \in [a,b]}$.

A version of the Gaussian kinematic formula for such a process would look at the expectations

$$\mathbb{E} \left\{ \mathcal{L}_j \left(\{ t \in M : f_t \in D \} \right) \right\}, \qquad (4.10.1)$$

where D is now a nice set in the function space in which the $f_t(\cdot)$ take their values. Although, in this setting, D will typically be infinite dimensional, by the principle of generalisation by mathematical optimism the corresponding Gaussian kinematic formula (4.6.11) should look the same.

In fact, it does not. However, it is no longer clear what the $\mathcal{L}_j(M)$ and the $\mathcal{M}_j^\gamma(D)$ are.

If f is, in an appropriate sense, stationary and isotropic, then by Hadwiger's theorem (cf. (1.2.14)) the $\mathcal{L}_j(M)$ should be the usual, Euclidean Lipschitz-Killing curvatures. Otherwise, they are going to have to be, once again, determined by a Riemannian metric, this time involving (Malliavin) derivatives of f.

However, even in the stationary, isotropic case, the meaning of the $\mathcal{M}_j^\gamma(D)$ is still not clear. However, as in the finite dimensional case, there exists a tube formula for infinite dimensional Gaussian measures, and this can be used the define corresponding Gauss Minkowski functionals. In many ways, this formula is the natural extension of Weyl's tube formula to infinite dimensions, and so of intrinsic geometric interest. Additional arguments also show that if μ is an additive functional on sets in $\mathbb{R}^\infty$ such that for any $A \in \mathbb{R}^N$, N finite, μ is rotation invariant, then μ has a representation as a sum of $\mathcal{M}_j$'s. This gives an infinite dimensional version of Hadwiger's theorem.

The details are far from trivial, but make for fascinating reading. See [82] for more.

Chapter 5
On Applications: Topological Inference

One of the most rewarding aspects of working on random field geometry is that one not only gets to do nice mathematics (an adjective with which we assume you agree if you have got this far in the notes) but one also gets to see the theory applied. Furthermore, the time delay from theory to applications is often measured in months, rather than the years or decades that typically link theory and practice. This close connection works in both directions: As new theory leads to new applications, new subject matter needs lead to the development of additional theory.

It is not possible to talk about this interaction without mentioning the late Keith Worsley, and his unique and fundamental contributions to the theory and applications of random fields.

Keith was primarily a biostatistican, and devoted the last two decades of his life to the development of methodologies for the statistical analysis of brain images. On the one hand, he worked closely with the medical imaging community, and on the other he worked with probabilists, applying their results in statistical practice, adapting them when necessary, and proving his own when the theoreticians were too slow for him. His unique ability to cross disciplines, and his effusive personality which enabled him to bring together people of very different backgrounds and interests, is sorely missed.

This chapter, since it is probably the one he would most have enjoyed reading, is dedicated to his memory.

In it we plan to give brief descriptions of a small number of applications of the theory of the preceeding chapters. Many of these have to do with thresholding issues, others with specific problems of statistical inference. Since both of these classes of applications rely on topological formulae they can be generically described as examples of *topological inference*, a term which, to the best of our knowledge, was first coined by Karl Friston. As we shall see in Chap. 6 this concept still covers many unexplored areas.

The applications we shall consider come from the areas of brain imaging and cosmology, but there are many more. As one might imagine, oceanography is rich in examples of time-space random fields, and the kind of theory that these notes contain has been applied there very richly. Indeed, much of it began there, with the fundamental papers of Longuet-Higgins (e.g. [63, 64])

R.J. Adler and J.E. Taylor, *Topological Complexity of Smooth Random Functions*, Lecture Notes in Mathematics 2019, DOI 10.1007/978-3-642-19580-8_5, © Springer-Verlag Berlin Heidelberg 2011

which gave very precise formulae for the expected number of critical and other specular of isotropic Gaussian fields in two and three dimensions. For a recent, very applied, paper looking at oceanographic data from a geometric viewpoint see [36]. There is also a wealth of additional applications in many areas of physics, including using Gaussian random field geometry to understand quantum chaos. A useful recent survey from the point of view of physics, with many useful references, is [28].

Meanwhile, the examples in this chapter should give you a limited idea of what can be done with topological inference. Hopefully, we shall one day finish *ARFG*, which will give a much fuller overview of applications, and also of Keith's contributions.

5.1 Local Structure of Extrema and the Euler Characteristic Heuristic

One of the main applications of the Gaussian kinematic formula has been its application in what has often been called the *Euler characteristic heuristic*. For many Gaussian processes, this is no longer a heuristic, because of the following theorem[1] in which, as usual, Euler characteristics are denoted by φ.

Theorem 5.1.1. *Let M and f be as in Theorem 4.8.1. Then there exists a constant $\sigma_c^2 > 0$, dependent on the distribution of f and the geometry of M, such that*

$$\liminf_{u \to \infty} -u^{-2} \log \left| \mathbb{E} \left\{ \varphi \left(A_u(f, M) \right) \right\} - \mathbb{P} \left\{ \sup_{t \in M} f(t) \geq u \right\} \right| \geq \frac{1}{2} \left(1 + \frac{1}{\sigma_c^2} \right).$$
(5.1.1)

Note that the importance of this result lies in the fact that while there are no known expressions for the exceedence probabilities $\mathbb{P}\{\sup_M f \geq u\}$, we have worked hard to develop explicit expressions for the mean Euler characteristic $\mathbb{E}\{A_u(f, M)\}$.

A little less formally, (5.1.1) states that there exist constants C and $\alpha = 1 + \sigma_c^2 > 1$ such that, for large enough u,

---

[1] For more details on this result see, for example, Chap. 14 of [8]. There you will also find information on the term σ_c^2. For example, if f is isotropic on $\mathbb{R}^N$, with unit variance and second spectral moments and monotonic covariance, then

$$\sigma_c^2 = \text{Var}\left(\frac{\partial^2 f(t)}{\partial t_1^2} \Big| f(t) \right) = \frac{\partial^4 \rho(t)}{\partial t_1^4} \Big|_{t=0} - 1.$$

That is, σ_c^2 is often explicitly computable. There is a long history of results of this kind, and you can find approaches somewhat different to that of [8] in [68] and [11].

$$\left| \mathbb{E}\left\{ \varphi\left(A_u(f,M)\right)\right\} - \mathbb{P}\left\{ \sup_{t\in M} f(t) \geq u \right\} \right| \leq C e^{-\alpha u^2/2}. \qquad (5.1.2)$$

The explicit expressions for $\mathbb{E}\left\{ \varphi\left(A_u(f,T)\right)\right\}$ developed in Chap. 4 show us that we can rewrite (5.1.2) as

$$\mathbb{P}\left\{ \sup_{t\in M} f(t) \geq u \right\} = C_0 \Psi(u) + e^{-u^2/2} \sum_{j=1}^{N} C_j u^{N-j} + o\left(e^{-\alpha u^2/2} \right), \qquad (5.1.3)$$

where the C_j are constants depending on the parameters of f and the geometry of M, and $N = \dim(M)$.

The final term in this expression is quite remarkable, for, if we think of the right hand side as an expansion of the form

$$C_0 \Psi(u) + e^{-u^2/2} \sum_{j=1}^{N} C_j u^{N-j} + \text{error}, \qquad (5.1.4)$$

it would be natural to expect that the error term here would be the 'next' term of what seems like the beginning of an infinite expansion for the exceedence probability, and so of order[2] $u^{-2} e^{-u^2/2}$. However (5.1.4) indicates that this is *not* the case. Since $\alpha > 1$, the error is actually *exponentially smaller* than this. Hence one can expect this approximation to work well in practice, as in fact it does.

So far, everything we have said in this chapter, bar, perhaps, the previous sentence, is rigorous. What, then, is the Euler characteristic *heuristic*? The heuristic lies in replacing exceedence probabilities by mean Euler characteristics, even for fields for which there is no rigorous parallel to Theorem 5.1.1, and, in all cases, without really knowing how large the level u has to be.

Experience, and simulations, have shown that the Euler characteristic heuristics works remarkably well. In the Gaussian case, $u = 2\sigma$ seems to be close enough to infinity for the mean Euler characteristic to approximate the true exceedence probability with an error of no more than .005, or relative error of no more than 10%. What is more important, is that the heuristic seems to work well even for non-Gaussian fields, even though we are still lacking a rigorous theorem which proves that this should be the case.

To understand why[3] the heuristic works, we need to consider the behaviour of random fields at high levels. The path to this is via what are known as Palm measures which describe the structure of a random field conditional on the occurrence of some special event, such as a local maximum, or level crossing

---

[2] Note that $\Psi(u)$ itself is $O(u^{-1}e^{-u^2/2})$.

[3] Such understanding is quite different to proving things. The proof of Theorem 5.1.1 does *not* just tighten up the following argument. It is, unfortunately, long, dry, and technical.

of a particular kind, occurring at a chosen point. The resulting conditional fields are described by what are generally known as *Slepian models*, or *Slepian processes*, after their discoverer, David Slepian. (cf. [55, 75]) for Gaussian processes on $\mathbb{R}$ and Lindgren [62] for Gaussian fields.)

While we shall not go into details here[4] consider the following useful example. Suppose that f is a stationary, centered, unit variance, Gaussian field on $\mathbb{R}^N$, regular in the sense of Condition 4.5.2. As usual, Λ is the matrix of second spectral moments. Then, conditional on f having a local maximum of height u at the point $t = 0$, with probability approaching one as $u \to \infty$ it has the following representation in a neighbourhood of the origin:

$$f(t) = u - \frac{u}{2\sigma} t\Lambda t' + O(1). \tag{5.1.5}$$

Now argue as follows, all with 'high probability': At high levels, at least in the neighbourhood of a local maximum, (5.1.5) shows that Gaussian fields are approximately parabolic. Therefore, ignoring boundary effects, their high level excursion sets are made up of a number, N_u say, of approximately elliptic components, each of Euler characteristic one. Consequently, $\varphi(A_u)$ and N_u are roughly equivalent, as are their means. Now suppose that the level u is high enough that $\mathbb{P}\{\sup_M f \geq u\} \approx \mathbb{P}\{N_u \geq 1\}$ are small, and that $\mathbb{P}\{N_u \geq 2\}$ is negligible compared to both of these. This leads to

$$\mathbb{P}\left\{\sup_{t \in M} f(t) \geq u\right\} \approx \mathbb{P}\left\{N_u \geq 1\right\}$$
$$\approx \mathbb{E}\left\{N_u\right\}$$
$$\approx \mathbb{E}\left\{\varphi(A_u)\right\}. \tag{5.1.6}$$

It is precisely this sequence of approximations that is the Euler characteristic heuristic, and the basis of the argument lies on two assumptions:

(a) Above high levels, the structure of the random field is simple enough that components of the excursion set with have Euler characteristic one with high probability.

(b) The probability that the random field will exceed a high level in two or more disjoint regions is negligible in comparison to the probability that it will do it once.

There are many variations of (5.1.5) for non-Gaussian fields. Typically, they are not as simple as in the Gaussian case, but it still follows that the individual connected components of the excursion set are geometrically simple. One can also convincingly argue that (b) holds. ("If the level is high, then it must be hard to exceed it even in one region, and so this has small probability.

---

[4] See Chap. 6 of either [2] or, better still, [9] for more details.

To exceed it in two regions must have a probability that is the square of this (assuming appropriate covariance decay) and so of much smaller order.")

Consequently, it is reasonable, and practical, to argue that an approximation like (5.1.2) holds for a wide variety of random fields although, obviously, the precise form of the bound on the right hand side is only expected to hold in the Gaussian case. This argument is the Euler characteristic heuristic in full generality.

Of course, if the argument above is correct, there must be other random variables that can be approximated by the Euler characteristic of the excursion set A_u. Indeed, there are, and these include the number of critical points above u, the number of local maxima above u, and the number of connected components of A_u. In each of these cases it is impossible to obtain a closed form expression for their expectation, but, on the other hand, an approximation akin to (5.1.2) is either known, or assumed, to hold. See Chap. 5 of *ARFG* for more details. If you are computationally oriented, you may also be interested in [6], which develops tools for the efficient simulation of Gaussian fields at high levels.

For more details on similar heuristics see the references of Footnotes 1 and 4, and also look at David Aldous' superb book [10] on the *Poisson clumping heuristic.*

5.2 Gaussian Related Random Fields

So far, these notes have concentrated almost exclusively on Gaussian random fields. Despite the richness of the related theory, this is clearly not a class of fields sufficiently wide to cover all applications. That is why, as early as Sect. 1.1 (cf. Fig. 1.1.1) we defined the (real valued and typically non-Gaussian) random fields

$$g(t) = F(f(t)) = F(f_1(t), \ldots, f_k(t)), \tag{5.2.1}$$

with $f : M \to \mathbb{R}^k$ Gaussian, with independent, identically distributed components, and $F : \mathbb{R}^k \to \mathbb{R}$ smooth. We shall call random fields of this type *Gaussian related.*

Choosing $k = 1$ and $F(x) = x$ takes us back to the real valued Gaussian case, but other simple choices take us to a wide range of interesting random fields. For example, suppose that the f_j are centred and of unit variance and consider the following four choices for F, where in the third we set $k = n + m$.

$$\sum_{i=1}^{k} x_i^2, \qquad \frac{x_1\sqrt{k-1}}{(\sum_{i=2}^{k} x_i^2)^{1/2}}, \qquad \frac{m \sum_{i=1}^{n} x_i^2}{n \sum_{i=n+1}^{n+m} x_i^2}, \qquad \min_{1 \leq i \leq k} x_i.$$

The corresponding random fields are known as χ^2 fields with k degrees of freedom, Student's T field with $k-1$ degrees of freedom, the F field with n and m degrees of freedom, and the conjunction field. If you have any familiarity with basic Statistics you will know that the corresponding distributions are almost as fundamental to statistical theory as is the Gaussian distribution.

Actually, there is no reason to restrict ourselves to real valued Gaussian related fields, for, given a set $F_1, \ldots, F_d$ of nice functions on $\mathbb{R}^k$ we can, analogously to (5.2.1), define a random field $g : M \to \mathbb{R}^d$ by

$$g(t) = F(f(t)) = \Big(F_1\big(f_1(t), \ldots, f_k(t)\big), \ldots, F_d\big(f_1(t), \ldots, f_k(t)\big) \Big). \quad (5.2.2)$$

Up to regularity conditions, we can allow the F_i to be quite general, although the f_i are our usual i.i.d. regular Gaussian fields. Note that it is only a notational issue to have different F_i depending on different f_j's, should we so desire.

Now suppose we want to know establish a version of the Gaussian kinematic formula for g. In particular, let $D \subset \mathbb{R}^d$ be nice. What can we say about the mean value of the geometric characteristics of $M \cap g^{-1}(D)$? A lot, once we note that

$$\begin{aligned} M \cap g^{-1}(D) = \{t \in M : g(t) \in D\} &= \{t \in M : f(t) \in F^{-1}(D)\} \\ &= M \cap f^{-1}\big(F^{-1}(D)\big) \\ &\overset{\Delta}{=} M \cap f^{-1}(D'), \end{aligned}$$

and so, as long as D and F were nice, so that $D' = F^{-1}(D)$ is nice, we are back in the setting of the Gaussian kinematic formula, despite the fact that we started with only a Gaussian related random field. From this it follows that

$$\mathbb{E}\left\{ \mathcal{L}_i\big(M \cap g^{-1}(D)\big) \right\} = \sum_{j=0}^{\dim M - i} \begin{bmatrix} i+j \\ j \end{bmatrix} (2\pi)^{-j/2} \mathcal{L}_{i+j}(M) \, \mathcal{M}_j^\gamma(D'). \quad (5.2.3)$$

Thus, we have turned a problem about Gaussian related fields into one about Gaussian fields. What is particularly useful, is that the non-Gaussian nature of the problem *has not impacted at all on the Lipschitz-Killing curvatures*. That is, the Lipschitz-Killing curvatures appearing in (5.2.3) are those related to the individual f_j, and are unaffected by the non-Gaussian aspects of the problem. Thus, for example, if the underlying f_j are stationary, then the Lipschitz-Killing curvatures are given by (4.5.2). In fact, any of the Lipschitz-Killing curvature computations of Chap. 4 still hold.

All that remains is computing the Gaussian Minkowski functionals, and we shall show you how to do this for one simple example. Further examples can be found in *RFG* and *ARFG*. In all examples one needs to check that the set D' in (5.2.3) satisfies the necessary regularity conditions, but this is done on a case by case basis.

5.2.1 χ^2 Fields

The example that we shall look at is the real valued χ^2 field, with k degrees of freedom, and $D = [u, \infty)$, so that we are looking at simple excursion sets of g. Thus, $d = 1$, $F(x) = \|x\|^2$, and $D' = \{x \in \mathbb{R}^k : \|x\|^2 \geq u\}$.

Recall (3.5.5), which in the present case leads to

$$\mathcal{M}_j^\gamma(D') = (-1)^k \frac{d^j}{dx^j} \mathbb{P}\{Z_k \geq x\}\Big|_{x=\sqrt{u}},$$

where Z_k is distributed as the (positive) square root of a χ_k^2 random variable, and so has probability density

$$p_k(x) = \frac{1}{\Gamma(k/2)2^{(k-2)/2}} x^{k-1} e^{-x^2/2}.$$

Direct calculations, exploiting the basic property (2.2.7) of Hermite polynomials, show that

$$\frac{d^{j-1}p_k(x)}{dx^{j-1}} = \frac{1}{\Gamma(k/2)2^{(k-2)/2}} \sum_{i=0}^{j-1} \binom{j-1}{i}(-1)^i \frac{d^{j-1-i}x^{k-1}}{dx^{j-1-i}} H_i(x) e^{-x^2/2}$$

$$= \frac{e^{-x^2/2}}{\Gamma(k/2)2^{(k-2)/2}} \sum_{i=0}^{j-1} \binom{j-1}{i}(-1)^i \frac{d^{j-1-i}x^{k-1}}{dx^{j-1-i}} H_i(x).$$

The summation can be rewritten as

$$\sum_{i=0}^{j-1} \mathbb{1}_{\{k \geq j-i\}} \binom{j-1}{i}(-1)^i \frac{(k-1)!}{(k+i-j)!} x^{k+i-j} H_i(x)$$

$$= x^{k-j} \sum_{i=0}^{j-1} \sum_{l=0}^{\lfloor i/2 \rfloor} \mathbb{1}_{\{k \geq j-i\}} \binom{j-1}{i}(-1)^{i+l} \frac{(k-1)!}{(k+i-j)!} \frac{i!}{(i-2l)!l!2^l} x^{2i-2l}$$

$$= x^{k-j} \sum_{l=0}^{\lfloor \frac{j-1}{2} \rfloor} \sum_{i=2l}^{j-1} \mathbb{1}_{\{k \geq j-i\}} \binom{j-1}{i}(-1)^{i+l} \frac{(k-1)!}{(k+i-j)!} \frac{i!}{(i-2l)!l!2^l} x^{2i-2l}$$

$$= x^{k-j} \sum_{l=0}^{\lfloor \frac{j-1}{2} \rfloor} \sum_{m=0}^{j-1-2l} \mathbb{1}_{\{k \geq j-m-2l\}} \binom{k-1}{j-1-m-2l} \frac{(-1)^{m+l}(j-1)!}{m!l!2^l} x^{2m+2l}.$$

This immediately easily leads to the following expression for the $\mathcal{M}_j^\gamma(D')$ for $j \geq 1$:

$$\frac{u^{(k-j)/2}e^{-u/2}}{\Gamma(k/2)2^{(k-2)/2}} \sum_{l=0}^{\lfloor\frac{j-1}{2}\rfloor} \sum_{m=0}^{j-1-2l}$$
$$\times \mathbf{1}_{\{k \geq j-m-2l\}} \binom{k-1}{j-1-m-2l} \frac{(-1)^{j-1+m+l}(j-1)!}{m!l!2^l} u^{m+l}. \qquad (5.2.4)$$

When $j = 0$, $\mathcal{M}_0^\gamma(D')$ is simply $\mathbb{P}\left\{\chi_k^2 \geq u\right\}$.

Note that, having set up the general theory, all we needed to explicitly compute the mean Lipschitz-Killing curvatures for χ^2 random fields was some calculus. For many other Gaussian related fields the same is true, although the calculus may be a little more formidable. Nevertheless, it remains no more than calculus.

Another important point to note is that if you believe in the general applicability of the Euler characteristic, then, since we now have an explicit expression for the mean Euler characteristic of χ^2 excursion sets, we also have an approximation for χ^2 exceedence probabilties.

In the following sections we shall see how to use these results in practice.

5.3 Brain Imaging

Most of the material in this section comes from a popular review [88] by Keith Worsley which, although nowadays rather out of date, in terms of both statistical methods and imaging technology, still makes for an excellent introduction to topological inference in brain imaging. While the example is based on analysing Positron Emission Tomography (PET) data, the same principles hold for more sophisticated (e.g. fMRI) data.

A very recent review [56] by Friston and Kilner is an excellent place to start looking for more up to date references and techniques. It also treats the analysis of electroencephalographic (EEG) and magnetoencephalographic (MEG) data, which, unlike PET data, also has a time component to it.

One of the earliest experiments in brain imaging was conducted in 1990 at Montreal Neurological Institute (cf. [59]). In this experiment, subjects were injected with a radio isotope emitting positrons, which annihilate with nearby electrons to release gamma rays that are detected by PET. By careful reconstruction, it was possible to build up an image of blood flow in the brain, a measure of brain activity. This opened up the possibility of actually seeing which regions of the brain were activated by different stimuli, and so to actually see the brain 'thinking'.

In 1992, in one of the first experiments of its kind, (cf. [35]) subjects were told to perform a linguistic task, involving the silent reading of words on a screen, during the imaging process. By subtracting an image in which each subject was 'at rest' looking at a blank screen, the experimenters were able to see evidence of increased blood flow in certain brain regions corresponding to the effort required for the task.

The images were, however, very blurred, and the signal (if any) was very weak compared to the background noise, so to increase the signal-to-noise ratio the experiment was repeated on 10 subjects. The brain images were aligned in three dimensions, and the blood flow was averaged, leading to the results of Fig. 5.3.1.

In Fig. 5.3.1, the brain is rendered as a transparent solid with the rear left side facing the viewer. The ventricles in the center form a single connected hollow that gives the brain an Euler characteristic of 2. One of 80 slices through the brain is colour coded (red = high, purple = low) to show (a) average blood flow of 10 subjects under the rest condition and (b) under the task condition. The difference of the averaged blood flows, task − rest, is shown in (c). Note that although these figures show a continuous image, the raw data was actually stored as values at $128 \times 128 \times 80 = 1,310,720$ voxels. At each such voxel there 10 pairs of blood-flow values, one pair for each subject, one taken while the subject was performing the task, the other at rest.

The statistical problem is to decide whether the task has any affect on brain activity. To do this the standard deviation of the 10 differences (9 degrees of freedom) was computed, and is shown in (d). The Z statistic for testing for a significant increase in blood flow due to the task is (e), where at a voxel v the Z statistic is given by

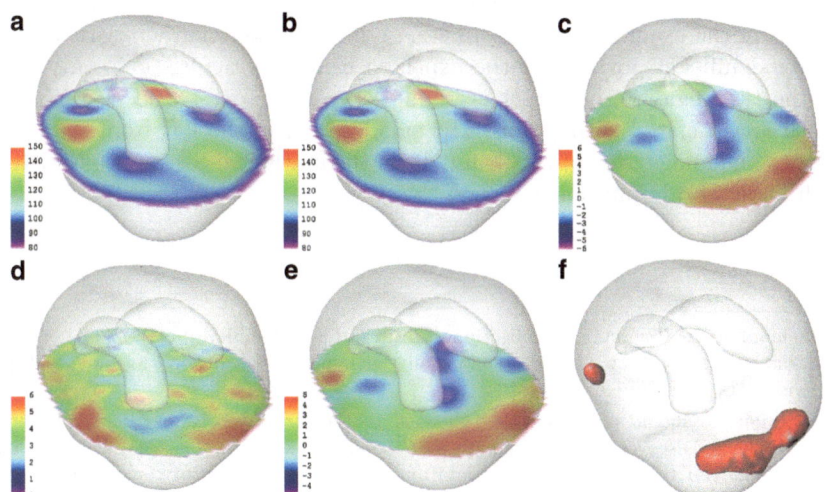

Fig. 5.3.1 A PET study showing regions of the brain activated by a reading task

$$Z(v) = \frac{\text{task at } v - \text{rest at } v}{\text{standard deviation at } v},$$

a ratio familiar to anyone who has taken even the most basic Statistics course.
We now have two tasks to perform:

(1) Decide whether or not performing the task has a (statistically) significant effect on the brain.
(2) If we decide that there is an effect, can we identify the region of the brain which is affected?

One way to approach (1) would be to perform independent normal difference tests at each of the 1,310,720 voxels, a patently ridiculous procedure since we would have no control over the Familywise Error Rate (FWER).[5] A now standard approach[6] to this problem is to assume that the random field Z observed on voxels is a sampled version of smooth random field defined over the brain, and that it fits into the theory of these notes.

The test statistic for the null hypothesis that there activation does not affect blood flow is then

$$z \overset{\Delta}{=} \sup_{v \in \text{Brain}} Z(v).$$

The tail distribution of t can then be computed, using the formula for the mean Euler characteristic and the Euler characteristic heuristic, and a rejection threshold[7] is determined.

Carrying out this procedure for the PET data gives a rejection threshold of $u = 4.22$ at the 5% level, and the red region in (f) of Fig. 5.3.1 shows those regions of the brain with Z above this level. The fact that there are two regions in the excursion set $A_{4.22}$ here, giving an Euler characteristic of 2, is highly significant, given that the expected Euler characteristic is only 0.05. Of the two regions, the larger is in the left visual cortex and extrastriate, which is associated with vision. The smaller one, in the left frontal lobe, is near the language area of the brain. Thus the statistical testing is consistent with known brain physiology.

---

[5] The familywise error rate of a collection of tests is the probability of making one or more false discoveries, or type I errors, among all the hypotheses when performing multiple pairwise tests.

[6] To be fair we should point out that there is another approach to error control, based on FDR (False Discovery Rate) techniques. We shall not describe this approach here, but rather point you to [12, 66] which explain and support FDR. We, of course, are partial to recent comparative studies [23, 24] which favour the random field approach described above. However, we shall leave it to subject matter specialists to fight over the details.

[7] A rejection threshold for t at level $\alpha \in [0, 1]$ is the maximum level u for which $\mathbb{P}\{t \geq u\} \geq \alpha$.

Of course, there is much missing and worrisome in the above analysis, but most of this has been fixed in the two decades since the above data were originally analysed.

For a start, it is hard to accept that the Z_v are normally distributed. After all, they are based on a sample of size 10, which is somewhat too small to appeal to the central limit theorem. However, it is reasonable to treat them as T random variables, with 9 degrees of freedom. However, T random variables, and T random fields, are Gaussian related, and so the arguments of the previous section, relying on the power of the Gaussian kinematic formula, enable us to repeat the above argument treating Z as a T, rather than Gaussian, random field. In this case there is no qualitative change in the results.

A relatively minor issue here is the passage from data on a discrete set of voxels to a continuous model. As we shall see, *en passant*, in Sect. 5.4 (cf. Fig. 5.4.1) this is not too serious a problem given the rather fine lattice on which the data are collected. However, it is an issue which requires addressing.[8]

Another problem lies in the fact that in our theory we have always assumed that the distributional parameters of our random fields were known, whereas here they need to be estimated. Although we did not state so explicitly, i = n the above analysis it was actually assumed that the field was isotropic,[9] so there was only one parameter to estimate, the second spectral moment λ_2. There are standard ways for doing this in the brain imaging community, so we shall not go into detail. However, it is important to note that once one is prepared to make assumptions of smoothness and isotropy, there is only one parameter that requires estimation.

Along the way one also needs to compute the Lipschitz-Killing curvatures of the brain. This, however, is easy and only needs to be done once. Given isotropy, the induced Riemannian metric on the brain is a scaled version of the standard Euclidean one, and so the Lipschitz-Killing curvatures are proportional to the brain's volume ($\mathcal{L}_3$), surface area ($2\mathcal{L}_2$), caliper diameter[10] ($\frac{1}{2}\mathcal{L}_1$), and Euler characteristic ($\mathcal{L}_0$).

Were we not to have assumed isotropy, nor even stationarity, the situation would have been much more complicated. In particular, we would have needed to estimate the Lipschitz-Killing curvatures in all their topological glory. We shall describe two ways to do this in the following section, along with two other examples coming from brain imaging problems.

---

[8] But not by us, and definitely not here.

[9] Actually, when the data was analysed, there was little choice but to assume isotropy, despite the fact the brain is definitely neither isotropic nor stationary. At that time, there was no Gaussian kinematic formula to use with the necessary milder assumptions.

[10] The caliper diameter of a set is defined by placing it between two parallel planes (or calipers), measuring the distance between the planes, and averaging over all rotations of the set.

5.4 Estimating Lipschitz-Killing Curvatures

Lipschitz-Killing curvatures have had a prominent role to play in almost everything we have treated so far, mainly due to the central place they take in the Gaussian kinematic formula. When we first met them in Sect. 1.2, we saw that they were reasonably easy to compute for convex sets in Euclidean spaces, simply by using Steiner's formula. In Sect. 3.4.2 we saw them defined in far greater generality, over stratified Riemannian manifolds, and made no attempt to discuss examples where they could be computed. The reason was that this is typically not easy to do.[11]

So what does a practicing statistician do, when there is a need to use the Gaussian kinematic formula for, say, a thresholding problem as in the previous section? There are essentially four practical approaches.

One approach relies on assuming that the Gaussian random fields underlying the problem are stationary and isotropic. In this case the Riemannian metric they induce on the parameter space (cf. Sect. 4.5) is Euclidean, and the Lipschitz-Killing curvatures $\mathcal{L}_k(M)$, as they appear in the Gaussian kinematic formula, are given by

$$\mathcal{L}_k(M) = \lambda_2^{k/2} \mathcal{L}_k^E(M),$$

where the $\mathcal{L}_k^E(M)$ are the Euclidean Lipschitz-Killing curvatures and λ_2 is the usual second spectral moment. Since the $\mathcal{L}_k^E(M)$ are generally computable from Steiner's formula (or Weyl's tube formula, if M is not convex) the main statistical problem reduces to estimating two parameters, λ_2 and the variance σ_2 (which in the theory we typically assumed was equal to 1). Estimating two parameters is not difficult, and there are various ways to do it. Typically, each subject area of application has its own favourites.

Another possibility is to acknowledge the possibility of non-stationarity, but to decide that if the application in mind is one of thresholding, it might suffice to calculate approximate expected Euler characteristics of excursion sets rather than precise ones. One way to do this is to take only the two highest order terms ($j = N = \dim M$ and $j = N - 1$) in the sum (4.8.3), which means that only $\mathcal{L}_N(M)$ and $\mathcal{L}_{N-1}(M)$ need to be computed.

It is not hard to show (cf. *RFG* for details) that, if M is a compact region in $\mathbb{R}^N$, then

$$\mathcal{L}_N(M) = \sigma^{-N} \int_M |\det \Lambda(t)|^{1/2} \, dt, \qquad (5.4.1)$$

where $\Lambda(t)$ is defined at (4.5.1).

---

[11] There is a nice example in Sect. 4.6.2 of *ARFG*, which treats a case in which the induced Riemannian metric arises from what is known as a 'scale-space' random field. But even this is not an easy calculation.

It is also possible to derive a reasonable expression for $\mathcal{L}_{N-1}(M)$, if M has C^2 boundary ∂M. This is given by

$$\mathcal{L}_{N-1}(M) = \frac{1}{2\sigma^{(N-1)}} \int_{\partial M} |\det \Lambda_{\partial T}(t)|^{1/2}\, \mathcal{H}_{N-1}(dt), \qquad (5.4.2)$$

where $\mathcal{H}_{N-1}$ is surface measure on ∂M. To define $\Lambda_{\partial T}(t)$, let $e_1(t), \ldots,$ $e_{N-1}(t)$ be an orthonormal basisfor the tangent space to M at $t \in \partial M$. Then, in analogy to (4.5.1), we define

$$(\Lambda_{\partial T}(t))_{ij} = \mathrm{Cov}\left(\frac{\partial f(t)}{\partial e_i(t)}, \frac{\partial f(t)}{\partial e_j(t)}\right).$$

There are situations in which these two top Lipschitz-Killing curvatures are not too hard to compute, despite a lack of stationarity or isotropy.

The third and fourth approaches arise when there is either no assumption of isotropy, or the parameter space is too complicated to make direct calculation of the $\mathcal{L}_k^E(M)$ analytically feasible. One of these takes a rather simple minded, but effective, regression approach. Details can be found in [5].

In the spirit of Sect. 5.2 let $f = F(g)$ be a Gaussian related random field, with everything nice enough that we can rewrite a special case of (5.2.3) as

$$\mathbb{E}\left\{\varphi\left(A_u(f, M)\right)\right\} = \sum_{k=0}^{\dim M} \mathcal{L}_k(M)\, \rho_k^F(u), \qquad (5.4.3)$$

for some (computable) functions ρ_k^F.

Now suppose that our data consists of a number of repeated observations f^i, $i = 1, \ldots, n$, of a random field f. We do not need to assume independence, but we do assume that the f^i have a common distribution. If, for a level u, a corresponding collection of excursion sets is given by

$$A_u^i = A_u\left(f^i, M\right),$$

a natural estimate of $\mathbb{E}\{\varphi(A_u(f, M))\}$ is given by the average

$$\frac{1}{n}\sum_{i=1}^{n} \varphi\left(A_u^i\right). \qquad (5.4.4)$$

Combining (5.4.3) and (5.4.4) yields the linear model

$$\frac{1}{n}\sum_{i=1}^{n} \varphi(A_u^i) = \sum_{k=0}^{\dim(M)} \mathcal{L}_k(M)\rho_k^F(u) + \varepsilon(u). \qquad (5.4.5)$$

Taking a sequence u_ℓ, $\ell = 1, \ldots, L$ of levels, we can assume[12] that the $\varepsilon(u_\ell)$ have a mean zero, multivariate, normal distribution. The covariances $\mathbb{E}\{\varepsilon(u_i)\varepsilon(u_j)\}$ are, of course, intractable.

Given (5.4.5), the $\mathcal{L}_k(M)$ can now be estimated via a generalized least squares regression, where GLS is required since the errors are both heteroskedastic and correlated. Carrying this out in practice is neither too hard nor trivial, and you can read about the details, which include optimal choice of the levels u_ℓ, in [5]. We shall show you some results of the procedure in a moment. However, even with the details, it is clear that a strong point of this approach is that it avoids the need to estimate of the underlying covariance, and actually does not even require a knowledge of what Lipschitz-Killing curvatures actually are. Furthermore, the procedure is essentially the same in any dimension.

A third approach to estimating Lipschitz-Killing curvatures is due to Taylor and Worsley [83] and is based on the idea of transforming non-stationary fields to local isotropy. As already discussed in Sect. 2.6.2, this is typically not possible. However, the Nash embedding theorem states that every Riemannian manifold can be isometrically embedded into some Euclidean space, where the meaning of isometric here is that the (Riemannian) geodesic distance between any two points in the original manifold is equal to the Euclidean distance in the embedded manifold. Using this, and assuming that the n replications of a field f are observed on points t_a, $a = 1, \ldots, A$, in M, the t_a are mapped into $S(\mathbb{R}^n)$ with the mapping

$$t_a \longrightarrow \frac{1}{\left(\sum_i (f^i(t_a))^2\right)^{1/2}} \left(f^1(t_a), \ldots, f^n(t_a)\right).$$

This creates a simplicial complex in $\mathbb{R}^n$ with A vertices, and it turns out that the Euclidean Lipschitz-Killing curvatures of this complex lead to good estimates of the Lipschitz-Killing curvatures that we require. Furthermore, the Lipschitz-Killing curvatures of the complex can be computed from local information on the sets of its vertices. The details are, again, neither too hard nor trivial, and you can find them in [83]. The advantage of this approach over the previous one seems to be that it attacks the $\mathcal{L}_j$ directly as geometric objects, rather than as coefficients in a regression equation. The disadvantage seems to be additional computation time, which one expects could become significant as the dimension of the problem grows.

How different are the previous two methods? In practice, it seems that although the approaches are very different, the results are similar. A simulation study was carried out in [5], involving 10,000 replications of systems of

---

[12] The assumption is justified either be appealing to a central limit theorem or the need to assume something tractable. In any case, the actual calculations for the model are L_2, so a Gaussian assumption is not really crucial here.

Gaussian random fields. The parameter spaces were squares, cubes, or two-dimensional spheres. The fields were simulated on grids, ranging from very coarse (5) to very fine (200). In the cube example, a 'grid size' of 100 means that the data was taken at $100 \times 100 \times 100$ points on a square lattice, which is the order of magnitude of the number of lattice points in an fMRI brain image. The covariance of the underlying random field was taken to match that of an fMRI example.

As well as giving estimates of the Lipschitz-Killing curvatures, the simulation study also used the Euler characteristic heuristic to estimate 95% threshold values, using the methods described in the previous section. Some of the results are summarised in Fig. 5.4.1. The plotted points in each box show average estimates of the 95% threshold as a function of grid size, with the smoothed curves based on logarithmic regression. The blue data is from the Adler–Bartz–Kou approach, and the pink is from the Taylor–Worsley approach. Note that in the fMRI range (cubic with grid ≈100) the thresholds are so close as to be visually indistinguishable. (In fact, the difference between them is less than .005.)

The vertical line at the top of the figure is the Euler characteristic heuristic approximation to the threshold, given the true parameters of the model, and is independent of the grid size. The lower, green curves are empirically observed threshold levels. The fact that these are lower than all the others is a consequence of the fact that the Euler characteristic heuristic approximation, whether it be based on known or estimated parameters, is related to the distribution of the supremum of a continuous random field observed over a continuous parameter space. The empirical thresholds, on the other hand, come from data collected on a discrete lattice.

As must be obvious, there is much, much more that needs to be said here. However, since our aim was merely to give you a flavour of brain imaging applications, we shall stop at this point.

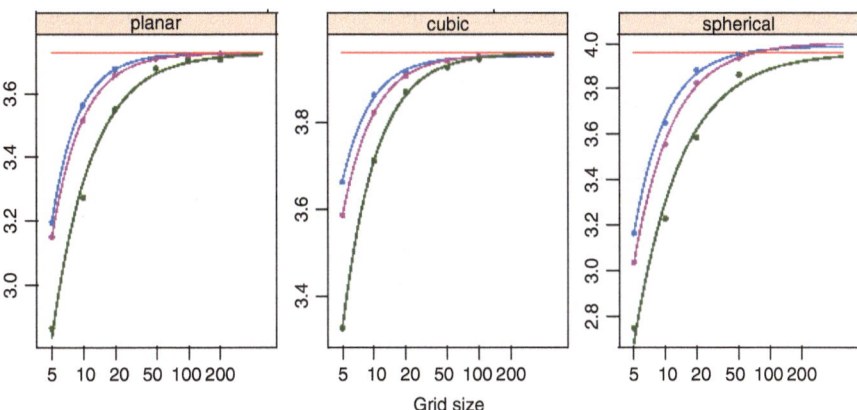

Fig. 5.4.1 Estimates of 95% thresholds for random fields, using the methods described in the text

5.5   Cosmology

We shall now look briefly at another two applications of the Gaussian kinematic formula, this time in the area of cosmology rather than brain imaging.
The difference will lie not so much in the subject matter as in the way the
theory is applied. In the brain imaging example the application we described
centered on the Euler characteristic heuristic, which was related to high level
excursion sets. In the cosmology applications, excursion sets at all levels have
a role to play.

We start with a some background, at the level of popular science.

In cosmology, the cosmic microwave background radiation (CMBR) is a
form of electromagnetic radiation filling the universe. CMBR is believed to
be radiation left over from an early stage in the development of the universe,
and its discovery is considered a landmark test of the Big Bang model of the
universe. When the universe was young it was smaller, hotter, and comprised
of a statistically uniform fog of hydrogen plasma. As the universe expanded,
both the plasma, and the radiation filling it, grew cooler. When the universe
cooled enough, stable atoms, which could no longer absorb thermal radiation,
began to form. The photons that were created during this cooling process have
been propagating ever since, though growing fainter and less energetic, and it
is these that are now measurable as CMBR. Its discovery in 1964 by American
radio astronomers Penzias and Wilson was the culmination of work initiated
in the 1940s, and earned them the 1978 Nobel prize. Its early measurement
and analysis in the 1980s won Mather and Smoot another Nobel prize in
2006.

Figure 5.5.1, taken from [44], shows what CMBR data looks like, after
much preprocessing. This data is from the Wilkinson Microwave Anisotropy
Probe (WMAP), a satellite which measures the CMBR across the full sky.
The parameter space is a sphere, since the probe collects only directional
data. The preprocessing involved in generating a picture like Fig. 5.5.1 is a
major task, but what is important for us is that, after the preprocessing, the

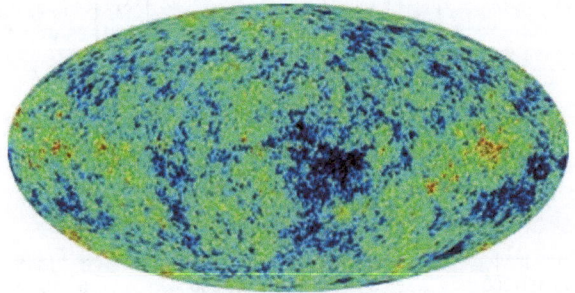

Fig. 5.5.1 WMAP 3-year data for the celestial sphere. The average temperature is plotted
as white. Higher temperatures are redder and cooler ones are bluer. For details see [44]

data are supposed to a realisation of a smooth, isotropic, Gaussian, random field on the sphere. The smoothness is a consequence of smoothing procedures in the preprocessing. The isotropy is a consequence of physical assumptions, and the Gaussianess a consequence of central limit theoretic arguments also at the Physics level.

Cosmologists are interested in using WMAP and similar data for testing the assumptions of Gaussianity and isotropy, and use the Gaussian kinematic formula to this end. The idea is simple: First compute the empirical Euler characteristic curve, $\varphi(A_u)$, over a range of u of physical interest, usually over a range of 6-7 standard deviations. Then estimate the parameters appearing in the GKF and, treating these as if they were the true parameters, compute the curve $\mathbb{E}\{\varphi(A_u)\}$. Alternatively, given the known general form of $\mathbb{E}\{\varphi(A_u)\}$, find the curve of this form that best fits the data. (This is closely related to the first method we described in the previous section for estimated Lipschitz-Killing curvatures.

One example is given in Fig. 5.5.2, in which the data points come from a particular frequency in the WMAP data, and the smooth curve is a best fit. To the uneducated eye, it would seem that the fit is excellent. Cosmologists, however, believe that the slight overshoot at level $\nu = 1$ is a significant indication of a physically important non-Gaussian perturbative phenomenon, and that studying it more than justifies the expenses involved in launching new satellites with higher resolution sensors.

This type of analysis is common throughout cosmology, and, as well as being used to compare data with theory, is also used to compare between competing theories, compare theory and data with simulation, etc. Many of these comparisons involve Gaussian vs. non-Gaussian assumptions, for which cases the fact that the Gaussian kinematic formula also gives us an explicit formula for $\mathbb{E}\{\varphi(A_u)\}$ in many non-Gaussian scenarios is particularly useful.

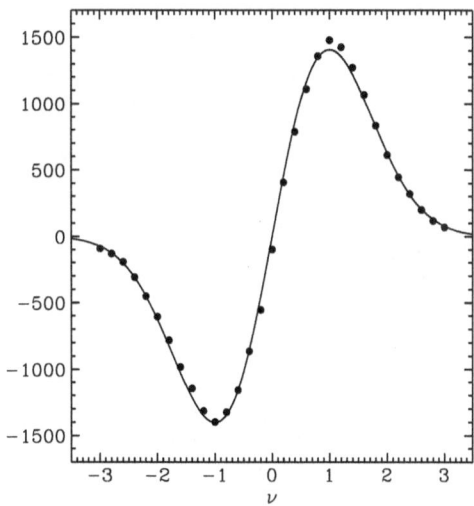

Fig. 5.5.2 Matching empirical Euler characteristics of excursion sets to a theoretical Gaussian expectation. For details see Fig. 4 of [44]

The analysis often also goes beyond the Euler characteristics of excursion sets, looking also at their Lipschitz-Killing curvatures. For some applications of this approach see, for example [65] and [72].

Another nice cosmological example comes from the statistical analysis of galactic density models. Here the most recent data comes from the Sloan Digital Sky Survey (SDSS) which is one of the most ambitious and influential surveys in the history of astronomy. A typical view of the data is given Fig. 5.5.3, which shows the results of three galactic surveys, results of which were published in 1986, 1994, and 2008. Each figure shows the excursion set A_u, where u is such that A_u covers 50% of the galactic volume.[13] The differences between the three surveys are primarily in terms of how large a slice of the universe has been surveyed. The fact that there is always a slice in the middle missing (the 'galactic plane') is a consequence of 'local' confounding effects related to the positioning of our own galaxy.

There are many competing theories of galactic structure, and many are tested by studying 'evolving universes' through N-body simulations. Roughly, what this means is that one starts at time zero with a N (recently of order 10^{10}) particles, spread out in space according to some random mechanism, and then allows them to move according to a physical model, often with a stochastic component of its own. Gravitational forces in the model eventually bring the particles together to form 'galaxies', and the resulting structure is

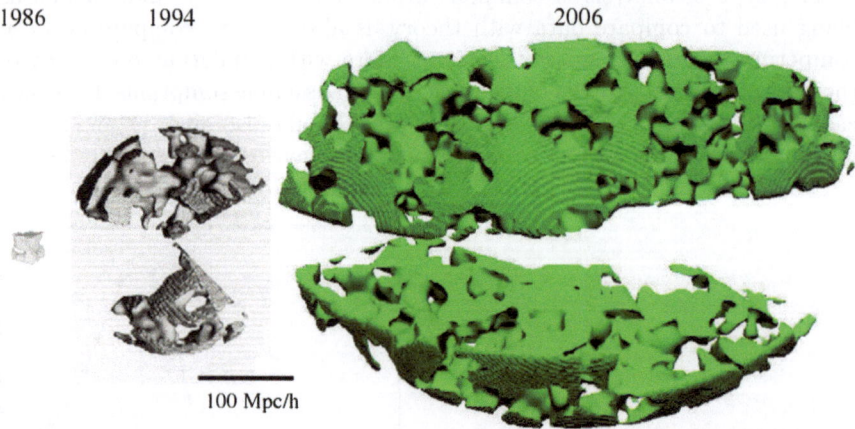

1986   1994   2006

100 Mpc/h

Fig. 5.5.3 Fifty percent high volume contours from three galactic surveys across three decades. From left to right, they are Gott et al. [45], Vogeley et al. [85], and Gott et al. [46]

---

[13] For those of you have read that galaxies actually have a fractal structure, the term 'galactic density' may sound a little strange. However, galactic density is effectively a smoothed, 'filtered' version of the matter spread throughout the visible universe. In terms of data, this is basically all that can be measured, and so fractal theories need to be converted to smoothed versions to allow comparison with data.

then smoothed to enable comparison with galactic survey data. Different distributions for the initial conditions and different physical models lead, naturally, to different 'final' galactic structures, each of which is actually a realisation of a smooth random field on a large subset of $\mathbb{R}^3$, nominally hundreds of megaparsecs in diameter.

Distinguishing between these theories is done much as in the CMBR case, except that now the random field is 3-dimensional. Nice, and quite typical, examples are given in Figs. 5.5.4 and 5.5.5.

Figure 5.5.4 shows five curves of the genus of excursion sets, genus being used more commonly than the Euler characteristic in this area. (Recall that the genus, $G(A)$, of a 3-dimensional set A is given by $G(A) = 1 - \varphi(A)$.) The blue curve (the highest at the zero level) gives the empirical genus of the excursion sets. The smooth curve gives the best Gaussian fit, and the other three are the results of N-body simulations of different kinds. The simple Gaussian actually seems to give a reasonable good fit to the SDSS data, although one of the simulations, the 'DH[14] mock SDSS', seems to do better. The MR mock SDSS, while doing badly in the centre of the range, seems to

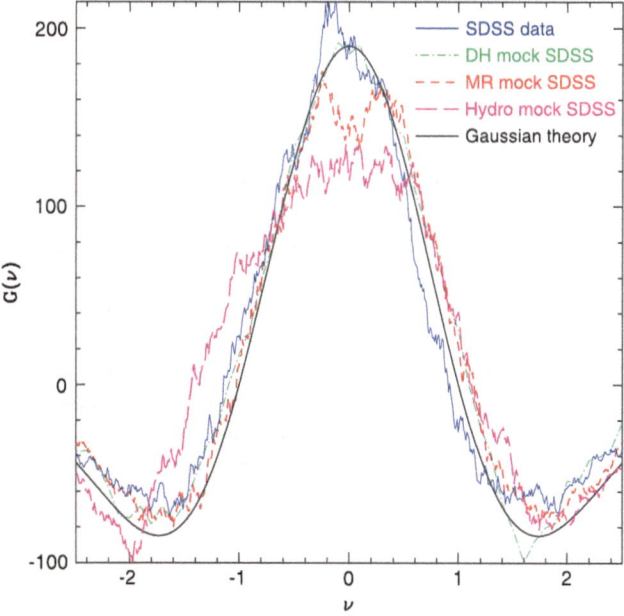

Fig. 5.5.4 Genus curves for the SDSS data, a Gaussian model, and individual simulations from three physical models

---

[14] DH relates to 'dark matter halo' and mock to the fact that the simulations produce fake, or mock, galaxies. MR refers to Millenium Run, an extremely large scale simulation. cf. [76].

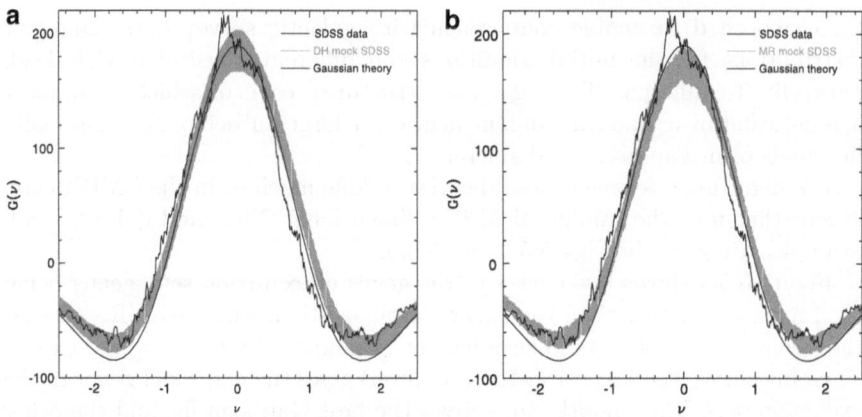

Fig. 5.5.5 Genus curves with shaded error regions for the (**a**) 100 DH and (**b**) 50 MR samples, compared with SDSS and a Gaussian model

do well in the tails. However, it is hard to tell from single simulations how good these fits are.

Figure 5.5.5, therefore, summarises the results of two larger scale simulations, and includes some error bars. The conclusion here seems to be that both the HD and MR models yield random fields of galactic density which, at least from the point of view of their excursion geometry, are very close to Gaussian in the centre of the distribution, but not in the tails. On the other hand, they fit the SDSS data in the tails, but not at the centre. This, interestingly, is the same region in which the CMBR data was not well s fitted by a Gaussian model, and thus the same issue of non-Gaussian perturbative effects arise here.

Hopefully, this chapter gave you some ideas as to how the information in the Gaussian kinematic formula can be exploited in applied scenarios. The applications are much broader than we have presented, as well as being deeper and more serious. Perhaps, one day, *ARFG* will be completed and you will have somewhere to turn for more details. In the meantime, there are very rich collections of papers out there which often make for fascinating reading.

Chapter 6
Algebraic Topology of Excursion Sets: A New Challenge

So far, when looking at the topology of excursion sets, these notes have primarily been concerned with differential topology. The main concepts which with we worked – Morse theory, Lipschitz-Killing curvatures, Gaussian Minkowski functionals – all come from this branch of topology.

Nevertheless, perhaps the most important of our tools, the Euler characteristic, is as much a concept of algebraic topology as it is of differential topology, and it is interesting to ask what might happen if we tried to push further into this realm. While algebraic topology is an area not commonly frequented by random fields researchers with a statistics or probability background, the aim of this brief chapter is to try to convince you that we have been missing out on an important, rich, and potentially very useful class of problems.

Algebraic topology is a branch of mathematics which uses tools from abstract algebra to study topological spaces. One of the main reasons for working in an algebraic world rather a geometric one is that while it becomes very hard, if not impossible, to picture geometry in dimensions greater than three the algebraic approach, at least in principle, works equally well in all dimensions. Thus, one wonders if it can help in the study of random field geometry.

Fortunately, over the last few years there has been a very interesting and rather exciting development in algebraic topology, as some of its practitioners have begun looking out beyond the inner beauty of their subject and seeing if they can apply it to problems in the 'real world'. As a result, 'applied algebraic topology' is no longer an oxymoron, and although it is true that at this point sophisticated applications are still few and far between, there is a growing feeling that the gap between theory and practice is closing. A very lively discussion of this trend can be found in Rob Ghrist's review [40], book in progress [41] and website on a project on sensor topology for minimal planning [42]. Gunnar Carlsson's webpage [19], which describes a large Stanford TDA (topological data analysis) project, and a DARPA webpage [25] describing a broad based project, also help explain the reasons why so many people have been so attracted to this direction.

R.J. Adler and J.E. Taylor, *Topological Complexity of Smooth Random Functions*, Lecture Notes in Mathematics 2019, DOI 10.1007/978-3-642-19580-8_6, © Springer-Verlag Berlin Heidelberg 2011

The aim of this chapter is to describe some of the new ideas that have arisen in applied algebraic topology and, in particular, to exploit some of them in the setting of random fields. It is, to a large extent, taken from the expository paper [7] where you can find more details and more examples.

6.1 Persistent Homology and Barcodes

In this section we are going to give a very brief and sketchy introduction to some basic notions of algebraic topology. A concise, yet very clear introduction to the topics that concern us can be found in [17,41], while [48,84] are good examples of a thorough coverage of homology theory. Recent excellent and quite different reviews by Carlsson [18,21], Edelsbrunner and Harer [34], and Ghrist [1,26,27,40] give a broad exposition of the basics of persistent homology.

Algebraic topology focuses on studying topology by assigning algebraic, group theoretic, structures to topological spaces M. Thus, homology, cohomology and homotopy groups can be used to classify objects into classes of 'similar shape'. We shall focus on homology. If M is of dimension N, then it has $N+1$ homology groups, each one of which is an abelian group. The zero-th homology $H_0(M)$ is generated by elements that represent connected components of M. For $k \geq 1$ the k-th homology group $H_k(M)$ is generated by elements representing k-dimensional 'loops' in M. The rank of $H_k(M)$, denoted by β_k, is called the k-th *Betti number*. For M compact and $k \geq 1$, β_k, measures the number of k-dimensional holes in M, while β_0 counts the number of connected components. It is a deep result, which we already met at (1.2.15), that our old friend the Euler characteristic is given by the alternating sum

$$\varphi(M) = \sum_{k=0}^{N} (-1)^k \beta_k(M). \tag{6.1.1}$$

To explain the idea of persistent homology, we shall work in the setting of the so-called 'Morse filtration' of excursion sets.

6.1.1 Barcodes of Excursion Sets

Suppose that M is a nice space, that $f : M \to \mathbb{R}$ is smooth, and consider the usual excursion sets $A_u = \{t \in M : f(t) \in [u, \infty)\}$. Note that if $u \geq v$ then $A_u \subseteq A_v$. Going from u to v, components of A_u may merge and new components may be born, and possibly later merge with one another or with the components of A_u. Similarly, the topology of these components may change, as holes and other structures form and disappear. Following the topology of these sets, as a function of u, by following their homology, is an example of persistent homology. The term 'persistence' comes from the fact that as the level u changes there is no change in homology until reaching a level u which is a critical point of f; i.e. the topology of the excursion sets remains static, or 'persists', between the heights of critical points. This basic observation is actually at the core of Morse theory.[1] However, the persistence of persistent homology goes further. For example, when two components merge, one treats the first of these to have appeared as if it is continuing its existence beyond point at which the merger occurs.

A useful way to describe persistent homology is via the notion of barcodes. Assuming that both f and M are smooth enough, then, if A_u is non-empty, $\dim(A_u)$ will typically be $N = \dim(M)$. A barcode for the excursion sets of f is then a collection of $N+1$ graphs, one for each collection of homology groups of common order. A bar in the k-th graph, starting at u_1 and ending at u_2 ($u_1 \geq u_2$) indicates the existence of a generator of $H_k(A_u)$ that first appeared at level u_1 and disappeared at level u_2. An example is given in Fig. 6.1.1, in which the function f is actually the realisation of a smooth random field on the unit square. The top seven boxes show there the surfaces generated by a 2-dimensional random field above excursion sets A_u for different levels u. To determine the level for each figure, follow the vertical line down to the scale at the bottom of the barcode. As the vertical lines pass through the boxes labelled H_0 and H_1, the number of intersections with bars in the H_0 (H_1) box gives the number of connected components (resp. holes) in A_u. Thus, at $u \sim 1.9$, A_u has 4 connected components but no holes, while at $u \sim -1.2$, A_u has only 1 connected component, but 9 holes. The horizontal lengths of the bars indicate how long the different topological structures (generators of the homology groups) persist.

Figure 6.1.2 is even more impressive, since it shows a three dimensional example. The barcode diagram is to be read as for Fig. 6.1.1, with two differences: The top 7 boxes now display the excursion sets themselves and the values of the field are colour coded. Furthermore, there are now three homology-groups/barcode-boxes, representing connected components, handles, and holes.

Note that, as opposed to the 2-dimensional case, it is almost impossible to say anything about topology just by looking at the boxes with the excursion sets at the top of the figure, but there is a lot of immediate visual information available in the barcodes. This phenomenon becomes even more marked as the dimension N of the parameter space increases. While it may be impossible

---

[1] In fact, the results of Sect. 3.8, which linked Euler characteristics of excursion sets to indices of critical points is a consequence of this observation, homology theory, and (6.1.1).

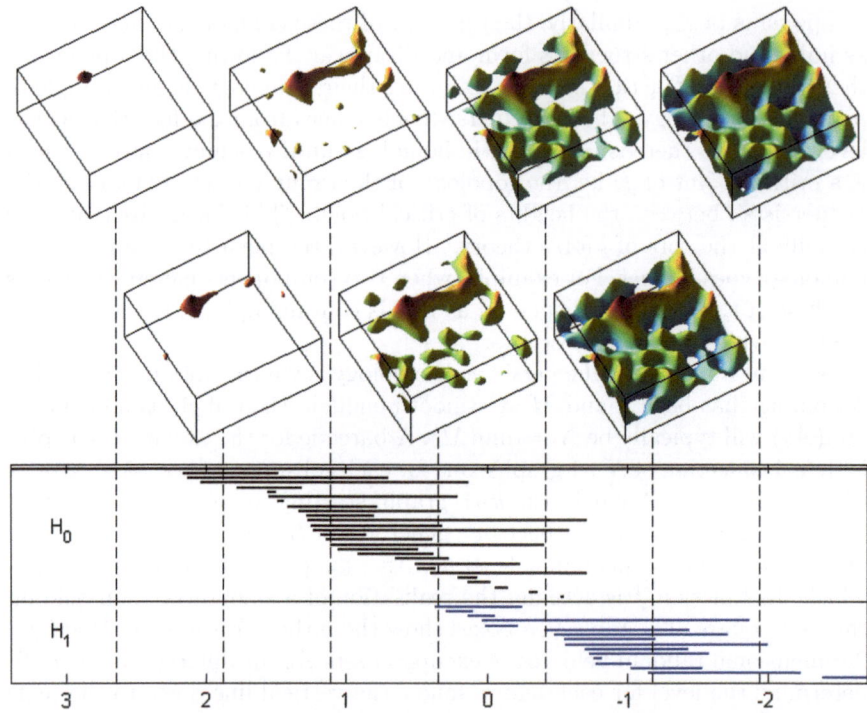

Fig. 6.1.1 Barcodes for the excursion sets of a function on $[0,1]^2$. Computation of the barcodes was carried out in Matlab using Plex (Persistent Homology Computations) from Stanford [20]. Note the unusual left to right ordering of the horizontal axis, with high levels on the right rather than the left

to imagine what a five dimensional excursion set looks like, it is easy to look at a barcode with six sets of bars for the six persistent homologies.

6.2 Barcode Distributions

A fascinating and challenging open area is determining distributional properties of bar codes. Note that one should treat the entire collection of bar codes generated by a random field as a single random variable, albeit taking values in a rather complicated space, viz. the space of barcodes. What can be said about the distribution of these barcodes? Where does one even start working on such a problem?

As usual, one starts with simulations. In particular, with 10,000 simulations of the random fields from which Figs. 6.1.1 and 6.1.2 were generated, giving two collections of 10,000 barcodes. In order to represent the barcode data in a reasonable fashion, it is convenient to employ *persistence diagrams*

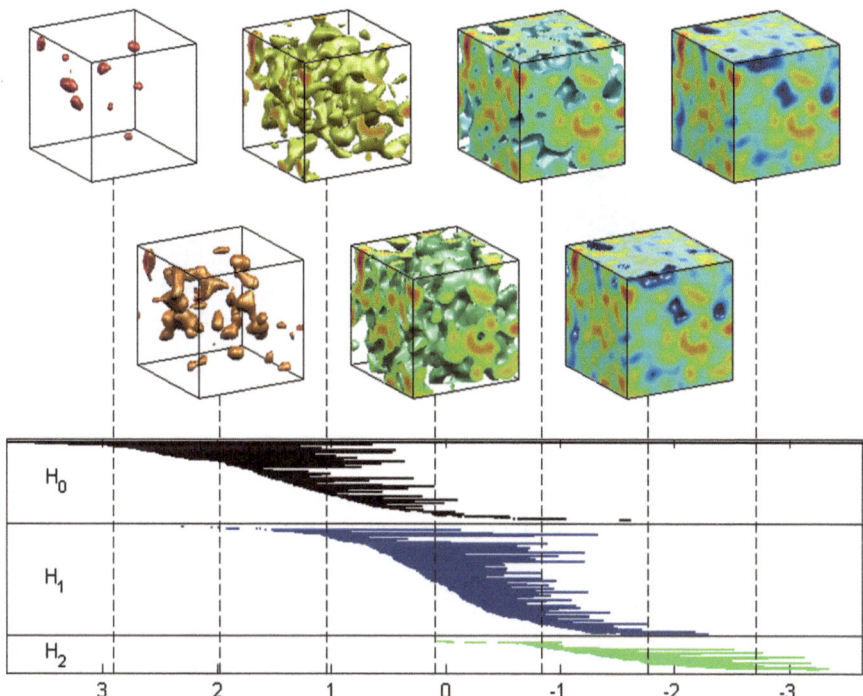

Fig. 6.1.2 Barcodes for the excursion sets of a 3-dimensional random field

rather than barcodes. To form a persistence diagram from the bars in H_k, one simply replaces each bar by a pair (x, y), where x is the level at which the bar begins and y the level at which it ends. Thus $x > y$ and the pair (x, y) lies in a half plane. In Fig. 6.2.1 the corresponding persistence diagrams for the complete simulation data are shown for H_0 and H_1.

Additional information on the barcodes is given in Fig. 6.2.2. What is shown there are the (marginal) distributions of the start and end points of the barcodes for H_0 and H_1 from the same simulation. A simple application of Morse theory, or, in this simple two dimensional setting, a little thought, leads to the realisation that the start points of the H_0 bars are all heights of local maxima of the field, while the end points of the H_1 bars correspond to local minima. These distributions have been well studied (although their precise form is not known) in the general theory of Gaussian random fields. The remaining start and end points correspond to different types of saddle points of the random field. However, what differentiates between the end point of a H_0 bar and the start point of a H_1 bar is global geometry and is not determined by the local behaviour of the field.

It would be interesting to know more about the real distributions underlying Figs. 6.2.1 and 6.2.2, but at this point we know very little.

There is, however, one thing that we do know.

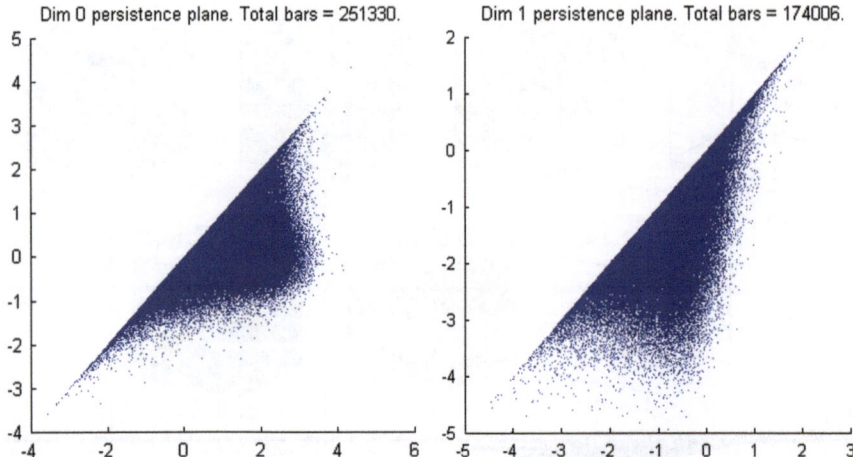

Fig. 6.2.1 Persistence diagrams for 10,000 simulations of an isotropic random field on the unit square. Note that the diagrams for H_0 and H_1 seem quite different

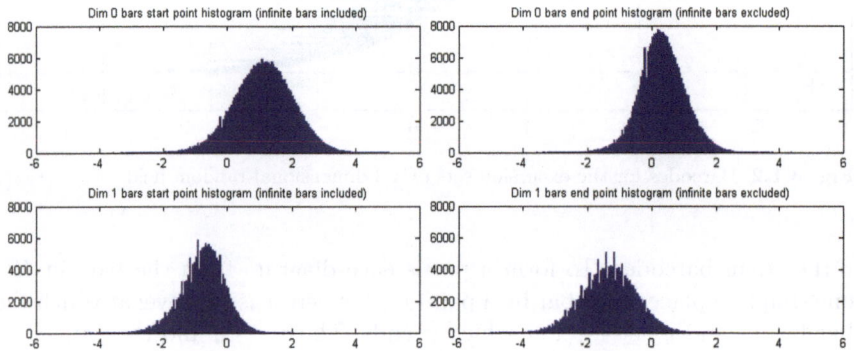

Fig. 6.2.2 Empirical distributions of start and end points of bars for the Gaussian field of Fig. 6.2.1

6.3 The Mean Euler Characteristic of the Barcodes of Gaussian Excursion Sets

We conclude these notes with the one existing theorem about Gaussian excursion set barcodes, for which we need a little notation, and a slight change of view. The latter is that instead of looking at excursion sets above levels, we shall look at excursions *below* levels, or sub-level sets rather than super-level sets. The difference is one of convention only, and will make it easier to compare the results of this section with the paper [14] from whence they came.

We denote a barcode diagram corresponding to the excursion sets $A_{(-\infty,u]}(f, M)$, $u \in \mathbb{R}$, by $\mathcal{B} = \mathcal{B}(f)$. Similarly, we denote a barcode diagram corresponding to the excursion sets $A_{(-\infty,u]}(f, M)$, $u \leq a$, by $\mathcal{B}_a = \mathcal{B}(f, a)$. We denote the individual bars in $\mathcal{B}$ (or $\mathcal{B}_a$) by b, their lengths by ℓ_b, and the degree of the homology group to which belongs the generator that they represent by $\mu(b)$. By $\mathcal{B}(f, a)$

We then define the Euler characteristic of a barcode $\mathcal{B}$ (or $\mathcal{B}_a$), after removing any bars of infinite length, to be

$$\chi(\mathcal{B}) \triangleq \sum_{b \in \mathcal{B}} (-1)^{\mu(b)} \ell(b).$$

The theorem is

Theorem 6.3.1. *As in the Gaussian kinematic formula, Theorem 4.8.1, $M \subset \mathbb{R}^N$ be a regular stratified manifold. Let $f = (f_1, \ldots, f_k) : M \to \mathbb{R}^k$ be a vector valued random process, the components of which are independent, identically distributed, real valued, Gaussian processes, regular in the sense of Condition 4.5.2, and with zero mean and constant unit variance. Assume that $F : \mathbb{R}^k \to \mathbb{R}$ is C^2 and set $g = F \circ f$. Then*

$$\mathbb{E}\left\{\chi(\mathcal{B}(g, g_{\max}))\right\} = \varphi(M)\left(\mathbb{E}\left\{g_{\max}\right\} - \mathbb{E}\{g\}\right)$$
$$+ \sum_{j=1}^{N} (2\pi)^{-j/2} \mathcal{L}_j(M) \int_{\mathbb{R}} \mathcal{M}_j(D_u) du,$$

where $D_u = F^{-1}((-\infty, u])$, $g_{\max} \triangleq \sup_{t \in M} g(t)$, and $\mathbb{E}\{g\} \triangleq \mathbb{E}\{g(t)\}$ for any $t \in M$.

If f is real, then

$$\mathbb{E}\{\chi(\mathcal{B}(f, a))\} = \varphi(M)\left(\phi(a) + a\Phi(a)\right) + \phi(a) \sum_{j=1}^{N} (2\pi)^{-j/2} \mathcal{L}_j(M) H_{j-2}(-a),$$

for any a. As usual, the $\mathcal{L}_j$, $j = 0, \ldots, N$ are the Lipschitz-Killing measures of M with respect to the metric induced by the f_i, and the $\mathcal{M}_j^\gamma$ are the Gaussian Minkowski functionals on $\mathbb{R}^k$. The functions ϕ and Φ are, respectively, the density and distribution functions of a standard Gaussian random variable.

Clearly, the content proof of Theorem 6.3.1 must somehow be related to the Gaussian kinematic formula, since it uses its conditions, and the form of the result is similar. For details, you will have to read [14], where you will also find a lot of other fascinating material. This includes an explanation of what is behind the result, and why it might be useful in applications.

One of our colleagues recently stated: "I can think of no two topics in mathematics further away from one another than probability and algebraic topology. There is probably no way to connect them."

Yet here, in Theorem 6.3.1, is an elegant connection, one of the first of its kind. Thus it is a tantalising way to end these notes, pointing, as it does, to an entire new class of problems and so the future, rather than the past.

References

1. A. Abrams and R. Ghrist. Finding topology in a factory: configuration spaces. *Amer. Math. Monthly*, 109(2):140–150, 2002.
2. R.J. Adler. *The Geometry of Random Fields*. John Wiley & Sons Ltd., Chichester, 1981. Reprinted in 2010 by SIAM, Classics of Applied Mathematics Series.
3. R.J. Adler. *An Introduction to Continuity, Extrema, and Related Topics for General Gaussian Processes*. Institute of Mathematical Statistics Lecture Notes—Monograph Series, 12. Institute of Mathematical Statistics, Hayward, CA, 1990.
4. R.J. Adler. On excursion sets, tube formulae, and maxima of random fields. *Annals of Applied Prob.*, 10:1–74, 2000.
5. R.J. Adler, K. Bartz, and S. Kou. Estimating metric curvatures for random fields. 2010. In preparation.
6. R.J. Adler, J.H. Blanchet, and J. Liu. Efficient Monte Carlo for high excursions of Gaussian random fields. 2010.
7. R.J. Adler, O. Bobrowski, M.S. Borman, E Subag, , and S. Weinberger. Persistent homology for random fields and complexes. In *Borrowing Strength: Theory Powering Applications, A Festschrift for Lawrence D. Brown*, pages 124–143. IMS Collections, 2010.
8. R.J. Adler and J.E. Taylor. *Random Fields and Geometry*. Springer, 2007.
9. R.J. Adler, J.E. Taylor, and K.J. Worsley. *Applications of Random Fields and Geometry: Foundations and Case Studies*. Springer-Verlag, 2011? In preparation, early chapters available at http://ie.technion.ac.il/~radler/publications.html.
10. D. Aldous. *Probability Approximations via the Poisson Clumping Heuristic*. Springer, New York, 1989.
11. J-M. Azaïs and M. Wschebor. *Level sets and extrema of random processes and fields*. John Wiley & Sons Inc., Hoboken, NJ, 2009.
12. R. Benjamini, Y. Heller. False discovery rates for spatial signals. *J. Amer. Statist. Assoc.*, 102(480):1272–1281, 2007.
13. A. Bernig and L. Bröcker. Lipschitz-Killing invariants. *Math. Nachr.*, 245:5–25, 2002.
14. O. Bobrowski and M.S. Borman. Euler integration of Gaussian random fields and persistent homology. 2010. In preparation.
15. V.I. Bogachev. *Gaussian Measures*, volume 62 of *Mathematical Surveys and Monographs*. American Mathematical Society, Providence, RI, 1998.
16. L. Bröcker and M. Kuppe. Integral geometry of tame sets. *Geom. Dedicata*, 82(1-3):285–323, 2000.
17. P. Bubenik and P.T. Kim. A statistical approach to persistent homology. *Homology, Homotopy and Applications*, 9(2):337–362, 2007.
18. G. Carlsson. Topology and data. *Bull. Amer. Math. Soc. (N.S.)*, 46(2):255–308, 2009.
19. G. Carlsson. Topological methods in scientific computing, statistics and computer science. 2010. comptop.stanford.edu/.

R.J. Adler and J.E. Taylor, *Topological Complexity of Smooth Random Functions*, Lecture Notes in Mathematics 2019, DOI 10.1007/978-3-642-19580-8, © Springer-Verlag Berlin Heidelberg 2011

20. G. Carlsson and V. de Silva. Plex: MATLAB software for computing persistent homology of finite simplicial complexes. comptop.stanford.edu/programs/ plex.
21. G. Carlsson and A. Zomorodian. The theory of multidimensional persistence. *Discrete Comput. Geom.*, 42(1):71–93, 2009.
22. S.S. Chern. On the kinematic formula in integral geometry. *J. Math. Mech.*, 16:101–118, 1966.
23. J. Chumbley, K. Worsley, G. Flandin, and K. Friston. Topological fdr for neuroimaging. *NeuroImage*, 49(4):3057–3064, 2010.
24. Justin R. Chumbley and Karl J. Friston. False discovery rate revisited: Fdr and topological inference using gaussian random fields. *NeuroImage*, 44(1):62–70, 2009.
25. DARPA. Topological data analysis. 2010. www.darpa.mil/dso/thrusts/math/ funmath/tda/index.htm.
26. V. de Silva and R. Ghrist. Coverage in sensor networks via persistent homology. *Algebr. Geom. Topol.*, 7:339–358, 2007.
27. V. de Silva and R. Ghrist. Homological sensor networks. *Notices Amer. Math. Soc.*, 54(1):10–17, 2007.
28. M.R. Dennis. Nodal densities of planar Gaussian random waves. *Eur. Phys. J.*, 145:191–210, 2007.
29. P. Diaconis and D. Freedman. A dozen de Finetti-style results in search of a theory. *Ann. Inst. H. Poincaré Probab. Statist.*, 23(2, suppl.):397–423, 1987.
30. P.W. Diaconis, M.L. Eaton, and S.L. Lauritzen. Finite de Finetti theorems in linear models and multivariate analysis. *Scand. J. Statist.*, 19(4):289–315, 1992.
31. R.M. Dudley. The sizes of compact subsets of hilbert space and continuity of Gaussian processes. *J. Funct. Anal.*, 1:290–330, 1967.
32. R.M. Dudley. Sample functions of the Gaussian process. *Ann. Probab.*, 1:66–103, 1973.
33. R.M. Dudley. *Uniform Central Limit Theorems*. Cambridge University Press, Cambridge, 1999.
34. H. Edelsbrunner and J. Harer. Persistent homology—a survey. In *Surveys on discrete and computational geometry*, volume 453 of *Contemp. Math.*, pages 257–282. Amer. Math. Soc., Providence, RI, 2008.
35. A. C. Evans, S. Marret, P. Neelin, L. Collins, K.J. Worsley, W. Dai, S. Milot, E. Meyer, and D. Bub. Anatomical mapping of functional activation in stereotactic coordinate space. *NeuroImage*, 1:43–53, 1992.
36. F. Fedele, G. Gallego, A. Benetazzo, A. Yezzi, and M.A. Tayfun. Euler characteristics and maxima of oceanic sea states. *IDRA08 31 Convegno Nazionale di Idraulica e Costruzioni Idrauliche Perugia*, pages 9–12, 2008.
37. H. Federer. Curvature measures. *Trans. Amer. Math. Soc.*, 93:418–491, 1959.
38. H. Federer. *Geometric Measure Theory*. Springer-Verlag, New York, 1969.
39. X. Fernique. *Fonctions Aléatoires Gaussiennes, Vecteurs Aléatoires Gaussiens*. Université de Montréal Centre de Recherches Mathématiques, Montreal, QC, 1997.
40. R. Ghrist. Barcodes: the persistent topology of data. *Bull. Amer. Math. Soc. (N.S.)*, 45(1):61–75 (electronic), 2008.
41. R. Ghrist. Topological methods in electrical and systems engineering, 2008. www.math.uiuc.edu/~ghrist/notes/appltop/.
42. R. Ghrist. Sensor Topology for Minimal Planning. 2009. www.darpa.mil/dso/ thrusts/math/funmath/stomp/index.htm.
43. M. Goresky and R. MacPherson. *Stratified Morse Theory*, volume 14 of *Ergebnisse der Mathematik und ihrer Grenzgebiete (3) [Results in Mathematics and Related Areas (3)]*. Springer-Verlag, Berlin, 1988.
44. J.R. Gott III, W.N. Colley, C.G. Park, C. Park, and C. Mugnolo. Genus topology of the cosmic microwave background from the WMAP 3-year data. *Monthly Notices of the Royal Astronomical Society*, 377(4):1668–1678, 2007.
45. J.R. Gott III, M. Dickinson, and AL Melott. The sponge-like topology of large-scale structure in the universe. *The Astrophysical Journal*, 306:341–357, 1986.

46. J.R. Gott III, D.C. Hambrick, M.S. Vogeley, J. Kim, C. Park, Y-Y. Choi, R. Cen, and K. Ostriker, J.P. Nagamine. Genus topology of structure in the sloan digital sky survey: Model testing. *The Astrophysical Journal*, 675:16–28, 2008.

47. H. Hadwiger. *Vorlesüngen Über Inhalt, Oberfläche und Isoperimetrie*. Springer-Verlag, Berlin, 1957.

48. A. Hatcher. *Algebraic Topology*. Cambridge University Press, 2002.

49. T. Hida and M. Hitsuda. *Gaussian Processes*. AMS, Providence, 1993.

50. H. Hotelling. Tubes and spheres in n-spaces and a class of statistical problems. *Amer. J. Math.*, 61:440–460, 1939.

51. D. Hug and R. Schneider. Kinematic and Crofton formulae of integral geometry: recent variants and extensions. In C. Barceló i Vidal, editor, *Homenatge al Professor Lus Santaló i Sors*, pages 51–80. Universitat de Girona, 2002.

52. S. Janson. *Gaussian Hilbert Spaces*. Cambridge University Press, Cambridge, 1997.

53. Søren Johansen and Iain M. Johnstone. Hotelling's theorem on the volume of tubes: some illustrations in simultaneous inference and data analysis. *Ann. Statist.*, 18(2):652–684, 1990.

54. M. Kac. On the average number of real roots of a random algebraic equation. *Bull. Amer. Math. Soc.*, 43:314–320, 1943.

55. M. Kac and D. Slepian. Large excursions of Gaussian processes. *Ann. Math. Statist.*, 30:1215–1228, 1959.

56. J.M Kilner and K.J. Friston. Topological inference for EEG and MEG data. *Ann. Appl. Stat.*, 4(3):1272–1290, 2010.

57. D.A. Klain and G-C. Rota. *Introduction to Geometric Probability*. Cambridge University Press, Cambridge, 1997.

58. M. Knowles and D Siegmund. On Hotelling's approach to testing for a nonlinear parameter in a regression. *Int. Statist. Rev*, 57:205–220, 1989.

59. R. Leblanc, E. Meyer, D. Bub, R. J. Zatorre, and A. C. Evans. Language localization with activation positron emission tomography scanning. *Neurosurgery*, 31(2):369–372, 1992.

60. M. Ledoux and M. Talagrand. *Probability in Banach Spaces. Isoperimetry and Processes*. Springer-Verlag, Berlin, 1991.

61. M.A. Lifshits. *Gaussian Random Functions*. Kluwer, Dordrecht, 1995.

62. G. Lindgren. Local maxima of Gaussian fields. *Ark. Mat.*, 10:195–218, 1972.

63. M.S. Longuet-Higgins. On the statistical distribution of the heights of sea waves. *J. Marine Res.*, 11:245–266, 1952.

64. M.S. Longuet-Higgins. The statistical analysis of a random moving surface. *Phil. Trans. Roy. Soc.*, A249:321–387, 1957.

65. P. Natoli, G. De Troia, C. Hikage, E. Komatsu, M. Migliaccio, PAR Ade, JJ Bock, JR Bond, J. Borrill, A. Boscaleri, et al. BOOMERanG constraints on primordial non-Gaussianity from analytical Minkowski functionals. *Monthly Notices of the Royal Astronomical Society*.

66. M. Perone-Pacifico, C. Genovese, I. Verdinelli, and L. Wasserman. False discovery control for random fields. *J. Amer. Statist. Assoc.*, 99(468):1002–1014, 2004.

67. M.J. Pflaum. *Analytic and Geometric Study of Stratified Spaces*, volume 1768 of *Lecture Notes in Mathematics*. Springer-Verlag, Berlin, 2001.

68. V.I. Piterbarg. *Asymptotic methods in the theory of Gaussian processes and fields*. American Mathematical Society, Providence, RI, 1996. Translated from the Russian by V. V. Piterbarg, Revised by the author.

69. S. O. Rice. The Distribution of the Maxima of a Random Curve. *Amer. J. Math.*, 61(2):409–416, 1939.

70. S.O. Rice. Mathematical analysis of random noise. *Bell System Tech. J.*, 24:46–156, 1945. Also in Wax, N. (Ed.) (1954), *Selected Papers on Noise and Stochastic Processes*, Dover, New York.

71. L.A. Santalo. *Integral Geometry and Geometric Probability*. Encyclopedia of Mathematics and its Applications. Addison-Wesley, Reading, 1976.

72. J. Schmalzing and K.M. Gorski. Minkowski functionals used in the morphological analysis of cosmic microwave background anisotropy maps. *Monthly Notices of the Royal Astronomical Society*, 297(2):355–365, 1998.

73. R. Schneider. *Convex Bodies: the Brunn-Minkowski Theory*, volume 44 of *Encyclopedia of Mathematics and its Applications*. Cambridge University Press, Cambridge, 1993.

74. D.O. Siegmund and K.J. Worsley. Testing for a signal with unknown location and scale in a stationary Gaussian random field. *Ann. Statist*, 23:608–639, 1995.

75. D. Slepian. On the zeros of Gaussian noise. In *Proc. Sympos. Time Series Analysis (Brown Univ., 1962)*, pages 104–115. Wiley, New York, 1963.

76. V. Springel, S.D.M. White, A. Jenkins, C.S. Frenk, N. Yoshida, L. Gao, J. Navarro, R. Thacker, D. Croton, J. Helly, et al. Simulations of the formation, evolution and clustering of galaxies and quasars. *Nature*, 435(7042):629–636, 2005.

77. Jiayang Sun. Tail probabilities of the maxima of Gaussian random fields. *Ann. Probab.*, 21(1):34–71, 1993.

78. A. Takemura and S. Kuriki. On the equivalence of the tube and Euler characteristic methods for the distribution of the maximum of Gaussian fields over piecewise smooth domains. *Ann. of Appl. Prob.*, 12(2):768–796, 2002.

79. J.E. Taylor. *Euler Characteristics for Gaussian Fields on Manifolds*. PhD thesis, McGill University, 2001.

80. J.E. Taylor. A Gaussian kinematic formula. *Ann. Probab.*, 34(1):122–158, 2006.

81. J.E. Taylor, A. Takemura, and R.J. Adler. Validity of the expected Euler characteristic heuristic. *Ann. Probab.*, 33(4):1362–1396, 2005.

82. J.E. Taylor and S. Vadlamani. Random fields and the geometry of Wiener space. 2010. In preparation.

83. J.E. Taylor and K.J. Worsley. Detecting sparse signals in random fields, with an application to brain mapping. *J. Amer. Statist. Assoc.*, 102(479):913–928, 2007.

84. J.W. Vick. *Homology Theory*. Academic Press New York, 1973.

85. M.S. Vogeley, C. Park, M.J. Geller, J.P. Huchin, and J.R. Gott. Topological analysis of the CfA redshift survey. *Astrophysical Journal*, 420:525–544, 1994.

86. H. Weyl. On the volume of tubes. *Amer. J. Math.*, 61:461–472, 1939.

87. K. J. Worsley. Local maxima and the expected Euler characteristic of excursion sets of χ^2, F and t fields. *Adv. in Appl. Probab.*, 26(1):13–42, 1994.

88. K. J. Worsley. The geometry of random images. *Chance*, 9(1):27–40, 1997.

89. K.J. Worsley. Boundary corrections for the expected Euler characteristic of excursion sets of random fields, with an application to astrophysics. *Adv. Appl. Probab.*, 27:943–959, 1995.

90. K.J. Worsley. Estimating the number of peaks in a random field using the hadwiger characteristic of excursion sets, with applications to medical images. *Ann. Statist.*, 23:640–669, 1995.

91. K.J. Worsley. Testing for signals with unknown location and scale in a χ^2 random field, with an application to fMRI. *Adv. in Appl. Probab.*, 33(4):773–793, 2001.

92. K.J. Worsley and K.J. Friston. A test for a conjunction. *Statist. Probab. Lett.*, 47(2):135–140, 2000.

93. M. Wschebor. *Surfaces aléatoires*, volume 1147 of *Lecture Notes in Mathematics*. Springer-Verlag, Berlin, 1985. Mesure géométrique des ensembles de niveau. [The geometric measure of level sets].

Notation Index

$A(f, M, u)$ Excursion set, 8
$B(\mathbb{R}^N)$ Unit ball in $\mathbb{R}^N$, 2
$B_\lambda(\mathbb{R}^N)$ Ball of radius λ in $\mathbb{R}^N$, 2
$B_d(t, \varepsilon)$ Ball in the canonical metric, 26
$C(m, i)$ Constants, 47
$C(s, t)$ Covariance function, 26
$d(s, t)$ Canonical metric, 26
$g,\ g_t$ Induced (Riemannian) metric, 74
$G_{n,\lambda}$ Isometry group on $S_\lambda(\mathbb{R}^n)$., 54
$H(M, d, \varepsilon)$ Log entropy function, 26
$H_n(x)$ Hermite polynomial, 17
I_N Unit cube in $\mathbb{R}^N$, 1
$N(M, d, \varepsilon)$ Metric entropy function, 26
$N_d(m, C)$, $N_d(m, \Sigma)$ Normal distribution, 13
N_u Upcrossings of the level u, 14
$R(X, Y)$ Riemannian curvature, 38
$R(X, Y, Z, W)$ Curvature tensor, 38
$S(\mathbb{R}^N)$ Unit sphere in $\mathbb{R}^N$, 1
$S(X, Y)$ Second fundamental form, 39
$S_\lambda(\mathbb{R}^N)$ Sphere of radius λ in $\mathbb{R}^N$, 1
$S_\nu(X, Y)$ Scalar second fundamental form, 39
s_n Area of sphere, 79
$T_t M$ Tangent space to M at t, 37

Subject Index

Lecture Notes in Mathematics

For information about earlier volumes
please contact your bookseller or Springer
LNM Online archive: springerlink.com

Vol. 1875: J. Pitman, Combinatorial Stochastic Processes. École d'Été de Probabilités de Saint-Flour XXXII – 2002. Editor: J. Picard (2006)

Vol. 1876: H. Herrlich, Axiom of Choice (2006)

Vol. 1877: J. Steuding, Value Distributions of L-Functions (2007)

Vol. 1878: R. Cerf, The Wulff Crystal in Ising and Percolation Models, École d'Été de Probabilités de Saint-Flour XXXIV – 2004. Editor: Jean Picard (2006)

Vol. 1879: G. Slade, The Lace Expansion and its Applications, École d'Été de Probabilités de Saint-Flour XXXIV – 2004. Editor: Jean Picard (2006)

Vol. 1880: S. Attal, A. Joye, C.-A. Pillet, Open Quantum Systems I, The Hamiltonian Approach (2006)

Vol. 1881: S. Attal, A. Joye, C.-A. Pillet, Open Quantum Systems II, The Markovian Approach (2006)

Vol. 1882: S. Attal, A. Joye, C.-A. Pillet, Open Quantum Systems III, Recent Developments (2006)

Vol. 1883: W. Van Assche, F. Marcellàn (Eds.), Orthogonal Polynomials and Special Functions, Computation and Application (2006)

Vol. 1884: N. Hayashi, E.I. Kaikina, P.I. Naumkin, I.A. Shishmarev, Asymptotics for Dissipative Nonlinear Equations (2006)

Vol. 1885: A. Telcs, The Art of Random Walks (2006)

Vol. 1886: S. Takamura, Splitting Deformations of Degenerations of Complex Curves (2006)

Vol. 1887: K. Habermann, L. Habermann, Introduction to Symplectic Dirac Operators (2006)

Vol. 1888: J. van der Hoeven, Transseries and Real Differential Algebra (2006)

Vol. 1889: G. Osipenko, Dynamical Systems, Graphs, and Algorithms (2006)

Vol. 1890: M. Bunge, J. Funk, Singular Coverings of Toposes (2006)

Vol. 1891: J.B. Friedlander, D.R. Heath-Brown, H. Iwaniec, J. Kaczorowski, Analytic Number Theory, Cetraro, Italy, 2002. Editors: A. Perelli, C. Viola (2006)

Vol. 1892: A. Baddeley, I. Bárány, R. Schneider, W. Weil, Stochastic Geometry, Martina Franca, Italy, 2004. Editor: W. Weil (2007)

Vol. 1893: H. Hanßmann, Local and Semi-Local Bifurcations in Hamiltonian Dynamical Systems, Results and Examples (2007)

Vol. 1894: C.W. Groetsch, Stable Approximate Evaluation of Unbounded Operators (2007)

Vol. 1895: L. Molnr, Selected Preserver Problems on Algebraic Structures of Linear Operators and on Function Spaces (2007)

Vol. 1896: P. Massart, Concentration Inequalities and Model Selection, École d'Été de Probabilités de Saint-Flour XXXIII – 2003. Editor: J. Picard (2007)

Vol. 1897: R. Doney, Fluctuation Theory for Lévy Processes, École d'Été de Probabilités de Saint-Flour XXXV – 2005. Editor: J. Picard (2007)

Vol. 1898: H.R. Beyer, Beyond Partial Differential Equations, On linear and Quasi-Linear Abstract Hyperbolic Evolution Equations (2007)

Vol. 1899: Séminaire de Probabilités XL. Editors: C. Donati-Martin, M. Émery, A. Rouault, C. Stricker (2007)

Vol. 1900: E. Bolthausen, A. Bovier (Eds.), Spin Glasses (2007)

Vol. 1901: O. Wittenberg, Intersections de deux quadriques et pinceaux de courbes de genre 1, Intersections of Two Quadrics and Pencils of Curves of Genus 1 (2007)

Vol. 1902: A. Isaev, Lectures on the Automorphism Groups of Kobayashi-Hyperbolic Manifolds (2007)

Vol. 1903: G. Kresin, V. Maz'ya, Sharp Real-Part Theorems (2007)

Vol. 1904: P. Giesl, Construction of Global Lyapunov Functions Using Radial Basis Functions (2007)

Vol. 1905: C. Prévôt, M. Röckner, A Concise Course on Stochastic Partial Differential Equations (2007)

Vol. 1906: T. Schuster, The Method of Approximate Inverse: Theory and Applications (2007)

Vol. 1907: M. Rasmussen, Attractivity and Bifurcation for Nonautonomous Dynamical Systems (2007)

Vol. 1908: T.J. Lyons, M. Caruana, T. Lévy, Differential Equations Driven by Rough Paths, Ecole d'Été de Probabilités de Saint-Flour XXXIV – 2004 (2007)

Vol. 1909: H. Akiyoshi, M. Sakuma, M. Wada, Y. Yamashita, Punctured Torus Groups and 2-Bridge Knot Groups (I) (2007)

Vol. 1910: V.D. Milman, G. Schechtman (Eds.), Geometric Aspects of Functional Analysis. Israel Seminar 2004-2005 (2007)

Vol. 1911: A. Bressan, D. Serre, M. Williams, K. Zumbrun, Hyperbolic Systems of Balance Laws. Cetraro, Italy 2003. Editor: P. Marcati (2007)

Vol. 1912: V. Berinde, Iterative Approximation of Fixed Points (2007)

Vol. 1913: J.E. Marsden, G. Misiołek, J.-P. Ortega, M. Perlmutter, T.S. Ratiu, Hamiltonian Reduction by Stages (2007)

Vol. 1914: G. Kutyniok, Affine Density in Wavelet Analysis (2007)

Vol. 1915: T. Bıyıkoğlu, J. Leydold, P.F. Stadler, Laplacian Eigenvectors of Graphs. Perron-Frobenius and Faber-Krahn Type Theorems (2007)

Vol. 1916: C. Villani, F. Rezakhanlou, Entropy Methods for the Boltzmann Equation. Editors: F. Golse, S. Olla (2008)

Vol. 1917: I. Veselić, Existence and Regularity Properties of the Integrated Density of States of Random Schrödinger (2008)

Vol. 1918: B. Roberts, R. Schmidt, Local Newforms for GSp(4) (2007)

Vol. 1919: R.A. Carmona, I. Ekeland, A. Kohatsu-Higa, J.-M. Lasry, P.-L. Lions, H. Pham, E. Taflin, Paris-Princeton Lectures on Mathematical Finance 2004. Editors: R.A. Carmona, E. Cinlar, I. Ekeland, E. Jouini, J.A. Scheinkman, N. Touzi (2007)

Vol. 1920: S.N. Evans, Probability and Real Trees. École d'Été de Probabilités de Saint-Flour XXXV – 2005 (2008)

Vol. 1921: J.P. Tian, Evolution Algebras and their Applications (2008)

Vol. 1922: A. Friedman (Ed.), Tutorials in Mathematical BioSciences IV. Evolution and Ecology (2008)

Vol. 1923: J.P.N. Bishwal, Parameter Estimation in Stochastic Differential Equations (2008)

Vol. 1924: M. Wilson, Littlewood-Paley Theory and Exponential-Square Integrability (2008)

Vol. 1925: M. du Sautoy, L. Woodward, Zeta Functions of Groups and Rings (2008)

Vol. 1926: L. Barreira, V. Claudia, Stability of Nonautonomous Differential Equations (2008)

Vol. 1927: L. Ambrosio, L. Caffarelli, M.G. Crandall, L.C. Evans, N. Fusco, Calculus of Variations and Non-Linear Partial Differential Equations. Cetraro, Italy 2005. Editors: B. Dacorogna, P. Marcellini (2008)

Vol. 1928: J. Jonsson, Simplicial Complexes of Graphs (2008)

Vol. 1929: Y. Mishura, Stochastic Calculus for Fractional Brownian Motion and Related Processes (2008)

Vol. 1930: J.M. Urbano, The Method of Intrinsic Scaling. A Systematic Approach to Regularity for Degenerate and Singular PDEs (2008)

Vol. 1931: M. Cowling, E. Frenkel, M. Kashiwara, A. Valette, D.A. Vogan, Jr., N.R. Wallach, Representation Theory and Complex Analysis. Venice, Italy 2004. Editors: E.C. Tarabusi, A. D'Agnolo, M. Picardello (2008)

Vol. 1932: A.A. Agrachev, A.S. Morse, E.D. Sontag, H.J. Sussmann, V.I. Utkin, Nonlinear and Optimal Control Theory. Cetraro, Italy 2004. Editors: P. Nistri, G. Stefani (2008)

Vol. 1933: M. Petkovic, Point Estimation of Root Finding Methods (2008)

Vol. 1934: C. Donati-Martin, M. Émery, A. Rouault, C. Stricker (Eds.), Séminaire de Probabilités XLI (2008)

Vol. 1935: A. Unterberger, Alternative Pseudodifferential Analysis (2008)

Vol. 1936: P. Magal, S. Ruan (Eds.), Structured Population Models in Biology and Epidemiology (2008)

Vol. 1937: G. Capriz, P. Giovine, P.M. Mariano (Eds.), Mathematical Models of Granular Matter (2008)

Vol. 1938: D. Auroux, F. Catanese, M. Manetti, P. Seidel, B. Siebert, I. Smith, G. Tian, Symplectic 4-Manifolds and Algebraic Surfaces. Cetraro, Italy 2003. Editors: F. Catanese, G. Tian (2008)

Vol. 1939: D. Boffi, F. Brezzi, L. Demkowicz, R.G. Durn, R.S. Falk, M. Fortin, Mixed Finite Elements, Compatibility Conditions, and Applications. Cetraro, Italy 2006. Editors: D. Boffi, L. Gastaldi (2008)

Vol. 1940: J. Banasiak, V. Capasso, M.A.J. Chaplain, M. Lachowicz, J. Mickisz, Multiscale Problems in the Life Sciences. From Microscopic to Macroscopic. Bedlewo, Poland 2006. Editors: V. Capasso, M. Lachowicz (2008)

Vol. 1941: S.M.J. Haran, Arithmetical Investigations. Representation Theory, Orthogonal Polynomials, and Quantum Interpolations (2008)

Vol. 1942: S. Albeverio, F. Flandoli, Y.G. Sinai, SPDE in Hydrodynamic. Recent Progress and Prospects. Cetraro, Italy 2005. Editors: G. Da Prato, M. Röckner (2008)

Vol. 1943: L.L. Bonilla (Ed.), Inverse Problems and Imaging. Martina Franca, Italy 2002 (2008)

Vol. 1944: A. Di Bartolo, G. Falcone, P. Plaumann, K. Strambach, Algebraic Groups and Lie Groups with Few Factors (2008)

Vol. 1945: F. Brauer, P. van den Driessche, J. Wu (Eds.), Mathematical Epidemiology (2008)

Vol. 1946: G. Allaire, A. Arnold, P. Degond, T.Y. Hou, Quantum Transport. Modelling, Analysis and Asymptotics. Cetraro, Italy 2006. Editors: N.B. Abdallah, G. Frosali (2008)

Vol. 1947: D. Abramovich, M. Mariño, M. Thaddeus, R. Vakil, Enumerative Invariants in Algebraic Geometry and String Theory. Cetraro, Italy 2005. Editors: K. Behrend, M. Manetti (2008)

Vol. 1948: F. Cao, J-L. Lisani, J-M. Morel, P. Musé, F. Sur, A Theory of Shape Identification (2008)

Vol. 1949: H.G. Feichtinger, B. Helffer, M.P. Lamoureux, N. Lerner, J. Toft, Pseudo-Differential Operators. Quantization and Signals. Cetraro, Italy 2006. Editors: L. Rodino, M.W. Wong (2008)

Vol. 1950: M. Bramson, Stability of Queueing Networks, École d'Été de Probabilités de Saint-Flour XXXVI - 2006 (2008)

Vol. 1951: A. Moltó, J. Orihuela, S. Troyanski, M. Valdivia, A Non Linear Transfer Technique for Renorming (2009)

Vol. 1952: R. Mikhailov, I.B.S. Passi, Lower Central and Dimension Series of Groups (2009)

Vol. 1953: K. Arwini, C.T.J. Dodson, Information Geometry (2008)

Vol. 1954: P. Biane, L. Bouten, F. Cipriani, N. Konno, N. Privault, Q. Xu, Quantum Potential Theory. Editors: U. Franz, M. Schuermann (2008)

Vol. 1955: M. Bernot, V. Caselles, J.-M. Morel, Optimal Transportation Networks (2008)

Vol. 1956: C.H. Chu, Matrix Convolution Operators on Groups (2008)

Vol. 1957: A. Guionnet, On Random Matrices: Macroscopic Asymptotics, École d'Été de Probabilités de Saint-Flour XXXVI – 2006 (2009)

Vol. 1958: M.C. Olsson, Compactifying Moduli Spaces for Abelian Varieties (2008)

Vol. 1959: Y. Nakkajima, A. Shiho, Weight Filtrations on Log Crystalline Cohomologies of Families of Open Smooth Varieties (2008)

Vol. 1960: J. Lipman, M. Hashimoto, Foundations of Grothendieck Duality for Diagrams of Schemes (2009)

Vol. 1961: G. Buttazzo, A. Pratelli, S. Solimini, E. Stepanov, Optimal Urban Networks via Mass Transportation (2009)

Vol. 1962: R. Dalang, D. Khoshnevisan, C. Mueller, D. Nualart, Y. Xiao, A Minicourse on Stochastic Partial Differential Equations (2009)

Vol. 1963: W. Siegert, Local Lyapunov Exponents (2009)

Vol. 1964: W. Roth, Operator-valued Measures and Integrals for Cone-valued Functions and Integrals for Cone-valued Functions (2009)

Vol. 1965: C. Chidume, Geometric Properties of Banach Spaces and Nonlinear Iterations (2009)

Vol. 1966: D. Deng, Y. Han, Harmonic Analysis on Spaces of Homogeneous Type (2009)

Vol. 1967: B. Fresse, Modules over Operads and Functors (2009)

Vol. 1968: R. Weissauer, Endoscopy for GSP(4) and the Cohomology of Siegel Modular Threefolds (2009)

Vol. 1969: B. Roynette, M. Yor, Penalising Brownian Paths (2009)

Vol. 1970: M. Biskup, A. Bovier, F. den Hollander, D. Ioffe, F. Martinelli, K. Netočný, F. Toninelli, Methods of Contemporary Mathematical Statistical Physics. Editor: R. Kotecký (2009)

Vol. 1971: L. Saint-Raymond, Hydrodynamic Limits of the Boltzmann Equation (2009)

Vol. 1972: T. Mochizuki, Donaldson Type Invariants for Algebraic Surfaces (2009)

Vol. 1973: M.A. Berger, L.H. Kauffmann, B. Khesin, H.K. Moffatt, R.L. Ricca, De W. Sumners, Lectures on Topological Fluid Mechanics. Cetraro, Italy 2001. Editor: R.L. Ricca (2009)

Vol. 1974: F. den Hollander, Random Polymers: École d'Été de Probabilités de Saint-Flour XXXVII – 2007 (2009)

Vol. 1975: J.C. Rohde, Cyclic Coverings, Calabi-Yau Manifolds and Complex Multiplication (2009)

Vol. 1976: N. Ginoux, The Dirac Spectrum (2009)

Vol. 1977: M.J. Gursky, E. Lanconelli, A. Malchiodi, G. Tarantello, X.-J. Wang, P.C. Yang, Geometric Analysis and PDEs. Cetraro, Italy 2001. Editors: A. Ambrosetti, S.-Y.A. Chang, A. Malchiodi (2009)

Recent Reprints and New Editions

LECTURE NOTES IN MATHEMATICS   🦄 **Springer**

Edited by J.-M. Morel, B. Teissier, P.K. Maini

Editorial Policy (for the publication of monographs)

1. Lecture Notes aim to report new developments in all areas of mathematics and their applications - quickly, informally and at a high level. Mathematical texts analysing new developments in modelling and numerical simulation are welcome.

   Monograph manuscripts should be reasonably self-contained and rounded off. Thus they may, and often will, present not only results of the author but also related work by other people. They may be based on specialised lecture courses. Furthermore, the manuscripts should provide sufficient motivation, examples and applications. This clearly distinguishes Lecture Notes from journal articles or technical reports which normally are very concise. Articles intended for a journal but too long to be accepted by most journals, usually do not have this "lecture notes" character. For similar reasons it is unusual for doctoral theses to be accepted for the Lecture Notes series, though habilitation theses may be appropriate.

2. Manuscripts should be submitted either online at www.editorialmanager.com/lnm to Springer's mathematics editorial in Heidelberg, or to one of the series editors. In general, manuscripts will be sent out to 2 external referees for evaluation. If a decision cannot yet be reached on the basis of the first 2 reports, further referees may be contacted: The author will be informed of this. A final decision to publish can be made only on the basis of the complete manuscript, however a refereeing process leading to a preliminary decision can be based on a pre-final or incomplete manuscript. The strict minimum amount of material that will be considered should include a detailed outline describing the planned contents of each chapter, a bibliography and several sample chapters.

   Authors should be aware that incomplete or insufficiently close to final manuscripts almost always result in longer refereeing times and nevertheless unclear referees' recommendations, making further refereeing of a final draft necessary.

   Authors should also be aware that parallel submission of their manuscript to another publisher while under consideration for LNM will in general lead to immediate rejection.

3. Manuscripts should in general be submitted in English. Final manuscripts should contain at least 100 pages of mathematical text and should always include

   - a table of contents;
   - an informative introduction, with adequate motivation and perhaps some historical remarks: it should be accessible to a reader not intimately familiar with the topic treated;
   - a subject index: as a rule this is genuinely helpful for the reader.

   For evaluation purposes, manuscripts may be submitted in print or electronic form (print form is still preferred by most referees), in the latter case preferably as pdf- or zipped ps-files. Lecture Notes volumes are, as a rule, printed digitally from the authors' files. To ensure best results, authors are asked to use the LaTeX2e style files available from Springer's web-server at:

   ftp://ftp.springer.de/pub/tex/latex/svmonot1/ (for monographs) and
   ftp://ftp.springer.de/pub/tex/latex/svmultt1/ (for summer schools/tutorials).
   Additional technical instructions, if necessary, are available on request from:
   lnm@springer.com.

4. Careful preparation of the manuscripts will help keep production time short besides ensuring satisfactory appearance of the finished book in print and online. After acceptance of the manuscript authors will be asked to prepare the final LaTeX source files and also the corresponding dvi-, pdf- or zipped ps-file. The LaTeX source files are essential for producing the full-text online version of the book (see
http://www.springerlink.com/openurl.asp?genre=journal&issn=0075-8434 for the existing online volumes of LNM).

   The actual production of a Lecture Notes volume takes approximately 12 weeks.

5. Authors receive a total of 50 free copies of their volume, but no royalties. They are entitled to a discount of 33.3% on the price of Springer books purchased for their personal use, if ordering directly from Springer.

6. Commitment to publish is made by letter of intent rather than by signing a formal contract. Springer-Verlag secures the copyright for each volume. Authors are free to reuse material contained in their LNM volumes in later publications: a brief written (or e-mail) request for formal permission is sufficient.

Addresses:
Professor J.-M. Morel, CMLA,
École Normale Supérieure de Cachan,
61 Avenue du Président Wilson, 94235 Cachan Cedex, France
E-mail: morel@cmla.ens-cachan.fr

Professor B. Teissier, Institut Mathématique de Jussieu,
UMR 7586 du CNRS, Équipe "Géométrie et Dynamique",
175 rue du Chevaleret,
75013 Paris, France
E-mail: teissier@math.jussieu.fr

For the "Mathematical Biosciences Subseries" of LNM:

Professor P.K. Maini, Center for Mathematical Biology,
Mathematical Institute, 24-29 St Giles,
Oxford OX1 3LP, UK
E-mail: maini@maths.ox.ac.uk

Springer, Mathematics Editorial, Tiergartenstr. 17,
69121 Heidelberg, Germany,
Tel.: +49 (6221) 487-259
Fax: +49 (6221) 4876-8259
E-mail: lnm@springer.com